THE ECOLOGY OF INTERCROPPING

John Vandermeer
Department of Biology, University of Michigan, USA

CAMBRIDGE UNIVERSITY PRESS
Cambridge, New York, Melbourne, Madrid, Cape Town, Singapore,
São Paulo, Delhi, Dubai, Tokyo, Mexico City

Cambridge University Press
The Edinburgh Building, Cambridge CB2 8RU, UK

Published in the United States of America by
Cambridge University Press, New York

www.cambridge.org
Information on this title: www.cambridge.org/9780521346894

First published 1989
First paperback edition 1992

A catalogue record for this publication is available from the British Library

Library of Congress Cataloguing in Publication Data

Vandermeer, John H.
The ecology of intercropping.
Bibliography: p.
Includes index.
1. Intercropping. 2. Agricultural ecology.
I. Title.
S603.5.V36 1989 631.5′8 87-33830

ISBN 978-0-521-34592-7 Hardback
ISBN 978-0-521-34689-4 Paperback

Contents

Preface

This book received its original impetus from two seemingly unrelated personal prejudices with which I found myself burdened in the late 1970s. First, the basic science of ecology seemed to be suffering from a lack of solid empiricism, its major tenets stemming from a combination of theory and speculation based on observed natural patterns. Second, the applied science of intercropping appeared to be void of a systematic theoretical framework within which the voluminous and usually empirical work might be interpreted. It seemed logical that the empiricism of intercropping might be usefully put to the service of ecology, whereas the theory of ecology might similarly form the basis of a framework for intercropping.

That ecology appeared to me as an overly speculative discipline may be little more than a reflection of Mario Bunge's observation that sciences tend to oscillate between periods of excessive observation – experimentation and excessive theory – speculation. During the 1940s, 1950s, and early 1960s the field of ecology seemed to be caught in a rather dull phase of observation and data collection. In the late 1960s ecologists discovered probabilities, matrices, and differential equations, and initiated a rich phase of theory development. In the 1980s a great deal of concern seems to be developing over the lack of a solid empirical base to the theoretical explosion, perhaps a new swing of the pendulum.

While it is certainly not true that all ecology is done in an empirical void, most ecologists would probably be forced to agree that theory and speculation have come out of balance with data. But the strong historical traditions of ecology indirectly mitigate against empiricism, in that ecologists seem to be either taxonomically oriented (bird ecologist, plant ecologist), or bound up to particular ecosystems, usually pristine ones. Theory thus has tended to evolve in response to the detailed observations of particular taxa or systems, and frequently both the theoretician and original observer have necessarily been satisfied with a theoretical framework which simply reflects natural patterns. Only rarely is the generation of a prediction distinctly deviating from the natural pattern regarded as a priority. This insistence on remaining within their own taxon or habitat has probably retarded the development of a solid empirical tradition in ecology. If one speculates that foraging height in a tree reflects competitive displacement which can in turn be shown theoretically to limit the number of species in the community, a certain false impression of empiricism can easily be generated. And if one insists on testing all predictions of the theory in the same community, empiricism will always be retarded. The sort of experimental manipulability required for serious empiricism is not likely to be found in a lake, or a forest, or among the birds in a savannah, or the mammals in a desert. Granted, the important observations and interesting speculations are dictated by these natural systems. But insisting on the location of the whole science within these inherently unmanipulable systems condemns ecology to remain in its quasi-empirical state. It is almost as though one were interested in elucidating the underlying principles of genetics, but insisted on working only on the genetics of elephants.

It is this hope for a solid empirical base to ecology that originally steered me towards agroecosystems in general. By their very nature agroecosystems are highly manipulable and thus amenable to the testing of ecological theory. The particular emphasis on intercrops stems from my interest in the tropics where intercrops are virtually the rule, at least on the lands of small producers.

A second personal prejudice, more recently developed, derives from reading the technical literature on intercropping. The technique itself has long been recognized as ubiquitous in tropical situations, and thus agricultural scientists have been studying it for quite some time. Much of this study focused on the very simple question: does the intercrop offer some sort of advantage over the associated monoculture? It is a question that is more difficult to answer than might be supposed initially, as discussed in Chapter 2. A more recent focus has been the more mechanistic question: if the intercrop is advantageous, why? It has been suggested that intercrops are advantageous because herbivores are deterred from finding their hosts, because nitrogen is more efficiently utilized, because evaporation is reduced, and a host of other mechanisms. While speculating and theorizing have become common in intercropping research, a general theoretical framework which might provide a logical ordering for all these questions has not yet emerged.

It had been the original intent of this book to demonstrate that on the one hand some simple principles from theoretical ecology, appropriately modified, could provide the needed theoretical framework which might make sense of the voluminous empirical work already accomplished, and perhaps help guide future empiricism in intercropping research. On the other hand, it seemed that the vast empirical base already developed in intercropping research might begin to provide an empirical foundation for much of ecology's theory. It was thus a dual purpose. To suggest to the ecologist that intercropping might be a rich laboratory for testing ecological theory and to suggest to the intercropping researcher that ecology might offer a framework within which an already strong empirical program might make more sense. If it serves only one of those purposes I will be satisfied.

But I should add that I had never intended a comprehensive review either of ecological theory or of intercropping research. The literature in ecological theory is immense and diverse, that on intercropping even larger. I have concentrated my treatment on those aspects of ecological theory that seem to me most likely to apply to the intercropping situation. Similarly, I have restricted myself in the intercropping literature to those examples that apply, sometimes somewhat tenuously, to the theory I present.

In most cases some jerryrigging of the original theory has been necessary. Sometimes the ecological form of the theory has a direct analog to intercropping problems. Thus, for example, the ecologists' qualitative dictum that two similar species will not coexist translates easily to a similar qualitative dictum for the intercropping researcher that two similar species will not overyield. But other times the ecological theory provides at best a weak analogy to the intercropping problem, and is perhaps only useful for the analytical techniques it suggests. Thus Levins' theory of fitness sets transforms into an analysis of facilitation in intercropping, where the invisible hand of evolution that chooses one genotype over the other is only weakly analogous to the farmer's decision as to what intercrop combination to use. But occasionally the ecological theory and its application to intercrops are perfectly homologous. The application of diffusion equations to herbivore movement should be the same in a tropical forest as in an intercrop. In short, I have borrowed and modified freely from the literature on ecological theory, trying to apply it to the major themes, as I see them, of intercropping research.

As usual this project has been a social event. Those whose constant interchanges probably qualify them for coauthorship are C. Bach, D. Boucher, W. Gamboa, M. Hansen, M. Liebman, H. McGuinness, A. Meyrat, I. Perfecto, M. Reeves, S. Risch, P. Rosset, B. Schultz and K. Yih. For reading and commenting on various sections of the manuscript I am indebted to C. Bach, D. Boucher, M. Liebman, I. Perfecto, A. Power, R. Rice, P. Rosset, K. Savoie, and E. Somarriba. A special thanks to D. Goldberg whose insightful comments caused me to completely reorganize the first half of the book. All of my colleagues in the New World Agriculture Group were a constant source of inspiration. Much of my initial work on intercrops was supported by the National Science Foundation (DEB8108271 and G-INT-8302848). Finally, this manuscript was prepared while I was a visiting professor at the Higher Institute of Agricultural Sciences (Instituto Superior de Ciencias Agropecuarias) in Managua, Nicaragua, supported by a Fullbright research fellowship. I am grateful to both institutions for their support and hospitality. I typed the manuscript myself.

1

Introduction: intercrops and ecology

Tropical travelers, from Darwin to my mother-in-law, will attest to the obvious fact that intercrops, two or more crops grown in association with one another, are common. My own travels in India, South-East Asia, and Latin America certainly offer no contradiction to this general rule. Quantitative estimates suggest that 98% of the cowpeas grown in Africa are intercropped (Arnon, 1972), 90% of the beans in Colombia are intercropped (Gutierrez *et al.*, 1975), and the percentage of cropped land in the tropics actually devoted to intercropping varies from a low of 17% for India (Srivastava, 1972) to a high of 94% in Malawi (Edje, 1979). Apparently, in El Salvador it used to be impossible to find sorghum *not* intercropped with maize (Alas, 1974, cited in Pinchinat *et al.*, 1976; Hawkins, 1984). Even in temperate North America, before the widespread use of modern varieties and mechanization, intercropping was apparently common (e.g. 57% of the soybean acreage in Ohio was grown in combination with maize in 1923 (Thatcher, 1925), and recently there seems to be increased interest in the subject, at least in the research community (e.g. Schultz *et al.*, 1982; Putnam *et al.*, 1985; Herbert *et al.*, 1984; Allen & Obura, 1983; Shackel & Hall, 1984). Thus, by any standard, intercropped agroecosystems are common. Their diversity and overall distribution are illustrated in a partial compilation of combinations that have been cited in the literature, as presented in Table 1.1. These 55 combinations are not in any way meant to be exhaustive, only to illustrate the commonness of the procedure.

Because the practise is so common it has attracted much attention, both from socially oriented investigators (anthropologists, rural sociologists, economists, etc.) and technically oriented investigators (agronomists, entomologists, ecologists, etc.). Reviews and bibliographies of intercropping and multiple cropping abound (e.g. Kass, 1978; Willey, 1979*a*; 1979*b*; 1981; Lamberts, 1980; IRRI, 1974:1975; Beets, 1982; Papendick, *et al.*, 1976; Sanchez, 1976; Francis, 1986b; Steiner, 1984; Govinden *et al.*, 1984). One might thus expect an excellent data base to have been accumulated along with something of a solid theoretical framework, given the ubiquity of the phenomenon, and quantity of scientific attention. But neither an excellent data

Table 1.1. *A few examples of intercropping combinations. All references to sorghum–pigeonpea and corn–bean intercrops have been ignored – they are too numerous to mention. Only one reference is given for each crop combination, not to imply that that reference is either the only or most important one, simply as an introduction to the literature for the interested reader. Most of the included combinations have undoubtedly been investigated numerous times other than the single reference given for each*

Crop combination	Region	Reference
Basic grains		
Maize and forage legumes	India	Singh & Chand, 1969
Jowar and soybean	India	Singh *et al.*, 1973
Sorghum and alfalfa	U.S.	Scott & Patterson, 1962
Sorghum and groundnuts	India	Bodade, 1964
Peas and oats	Germany	Gliemeroth, 1950
Sorghum and millet	Nigeria	Norman *et al.*, 1970
Oats and rye	U.S.	Pfahler, 1965
Maize and sorghum	El Salvador	Alas, 1974
Maize and sweetpotato	Philippines	Lawas, 1947
Maize and soybean	U.S.	Liboon & Harwood, 1975
Maize and cowpea	Mexico	Vandermeer *et al.*, 1983
Rice and tobacco	Taiwan	Sung & Wu, 1966
Oats and vetch	U.S.	Robinson, 1960
Maize and cassava	Philippines	Martinez, 1947
Jowar and arhar	India	Gupta, 1953
Maize and pigeonpea	India	Dalal, 1974
Rice and sugarbeet	China	Chang & Lin, 1960
Oats and barley	U.S.	Bussell, 1937
Jowar and groundnut	India	Bodade, 1964
Soybean and oats	U.S.	Brown & Graffis, 1976
Sorghum and oats	U.S.	
Wheat and oats	U.S.	Bailey, 1914
Millet and groundnut	India	Osiru & Kibira, 1979
Soybean and arhar	India	Sharma *et al.*, 1973
Maize and mungbeans	Philippines	Castin *et al.*, 1976
Maize and groundnut	Philippines	Punzalan, 1972
Soybean and rice	India	Reddy & Chatterjee, 1973
Wheat and ryegrass	Australia	Rerkasem, *et al.*, 1980
Oilcrops and sugarcane		
Cotton and groundnuts	India	Pillai *et al.*, 1957
Sugarcane and maize	India	Bhoj & Kapoor, 1970
Sugarcane and soybeans	Philippines	Paner, 1975

Table 1.1 (*cont.*)

Crop combination	Region	Reference
Sugarcane and sunflower	Mauritius	Rouillard & Mazery, 1969
Cotton and garlic	U.S.	Moursi, 1966
Cotton and sisal	Mozambique	Carvalho, 1969
Sugarcane and sweetpotato	Taiwan	Shia and Pao, 1964
Cotton and maize	Kenya	Grimes, 1948
Vegetables		
Tomatoes and watermelon	Nicaragua	Pers. obs.
Tomatoes and cucumbers	U.S.	Schultz *et al.*, 1982
Cabbage and tomatoes	Philippines	Begonia & Mercado, 1974
Tomatoes and soybean	U.S.	Vandermeer *et al.*, 1984
Vegetables and sugarcane	Puerto Rico	Lugo-Lopez, 1953
Radish and sunflower	England	Lakhani, 1976
Tomatoes and beans	Costa Rica	Rosset *et al.*, 1984
Perennials		
Rubber and coffee	Guatemala	Townsend *et al.*, 1964
Rubber and food crops	Indonesia	Senenayake, 1968
Coconuts and food crops	Sri Lanka	Santhirasegaram, 1967
Coconuts and cacao	Malaysia	Leach, 1971
Papaya and macadamia	Hawaii	Murashige, 1962
Bananas and coffee	Tanganyika	Robinson, 1962
Coconut and pineapple	Sri Lanka	Nair, 1983
Rubber and cacao	Costa Rica	Imle *et al.*, 1954
Olives and opuntia	Chile	Pers. obs.
Peach palms and cacao	Costa Rica	Pers. obs.
Jocote and pitya	Nicaragua	Pers. obs.
Rubber and cassava	Malaysia	Pushparajah & Tan, 1970

base nor a well-accepted theoretical framework has evolved. The phenomenon apparently involves more variables than initially come to mind and the agronomist based in the paradigm of more efficient use of nitrogen will possibly not notice the potential plant protection aspects so dominant on the mind of the entomologist, while the ecologist emphasizing soil conservation may be blind to the yield stability that draws the attention of the economist. In short, the practice of intercropping is far more complicated than monocultural production and has been resistent to the development of a central core of theory which might guide empirical work. Consequently, the voluminous empirical literature is eclectic, scattered, and sometimes confusing. Despite the excellent reviews cited above, the vast empirical base that has resulted from several decades of intercropping research has led to only modest gains in our ability to improve extant systems, devise new systems, or even evaluate unequivocally the efficacy of current practises. It would seem that the weak development of a core of theory is the problem. No matter how accurately we might measure the rate of fall of the apple, falling objects probably never would have made much sense if Newton's equations had not provided a theoretical framework within which those measurements could be interpreted.

My own prejudice is that the sort of theoretical framework that would be useful is similar to, if not identical with, that already developed in ecology. As I will argue below, many of the obvious questions that grow from intercropping research (and/or simple field observations of intercrops) have striking similarities with some of the classic questions of population and community ecology. Why not then try and apply the rich theoretical developments of ecology to the practice of intercropping, thus hopefully providing that theoretical framework within which empirical results might be more easily interpreted? Such is the rationale of this book.

The limited intentions of this book

We can begin by looking at what might be considered normal procedure in intercropping research. It has become something of a routine research procedure to ask why are intercrops so common? The answers to this question are as eclectic as intercrop combinations themselves. In one of the most complete reviews of the subject to date, Lamberts (1980) cites the following 'reasons' for intercropping:

1. increased productivity/yield advantages;
2. better use of available resources
   a. land, b. labor, c. time, d. water, e. nutrients;
3. reduction in damage caused by pests
   a. diseases, b. insects, c. weeds;
4. socio-economic and other advantages
   a. greater stability, b. economics, c. human nutrition, d. the 'biological aspect'.

While this list (along with Lambert's discussion of each topic) provides a comprehensive outline of those 'mechanisms' of intercrop advantage presented in the literature, it also demonstrates the ultimate complexity of understanding intercropping – from purely economic aspects, to sociological aspects, to agronomic aspects, to combinations of various disciplines. Under 'better usage of available resources', for example, we find land (basically a socio-political factor) along with labor (an economic factor), along with water and nutrients (agronomic factors).

The present work does not pretend to treat all of these factors. It specifically ignores all social, political, and most economic factors, and effectively restricts its attention to agronomic or ecological aspects. This approach in no way intends to suggest that those factors excluded are less important. On the contrary, Lambert's discussion leaves no doubt that the sociological component is at times overwhelmingly important. It is, nevertheless, necessary to take an exclusionary approach when we are at such a primitive level of understanding of the phenomenon at hand, as we are in the case of intercropping. Thus, what follows is the development of theoretical formulations aimed at understanding those aspects of intercropping that might be called 'strictly biological'.

The nature of ecological theory

There has been something of a controversy in ecology as to the nature of theory in general, and specifically what sort of theory would be appropriate for the admittedly complex subject of ecology. The literature on this subject alone is extensive (e.g. Levins, 1966, 1968; May, 1974, 1981; Simberloff, 1983; Roughgarden, 1983; Pielou, 1981; Levins & Lewontin, 1980).

We can conveniently summarize at least part of the debate as a conflict between two 'styles' of modeling: strategic versus tactical (May, 1974). Tactical models are designed to be used for a specific purpose and generally aim at a precise description of the phenomenon in question. Strategic models are designed to be very general representations of ecological phenomena, not necessarily precise descriptions thereof. Levins (1966) characterized these two extremes as simply different combinations of model characteristics. According to Levins, three of the most important characteristics of models are realism, generality and precision. It is difficult, if not impossible, to incorporate all three of these factors in a single theory, thus giving rise (usually) to models that are either realistic and precise but not general (tactical), or models that are realistic and general, but not precise (strategic). It is a continuum between a general-strategic 'abstract science' approach and a specific-tactical 'engineering' approach.

In this book we will be concerned with both types. While the main thrust is intended to be on the tactical engineering side, a solid strategic or 'abstract science' framework makes the development of tactical models so much easier, as many historical precedents tell us. What we desire is a general but realistic abstract framework which suggests various avenues for the development of tactical engineering methods. Thus, for example, Chapter 3 presents a very general (realistic and strategic) formulation of ecological competition theory applied to intercrops and then Chapter 10 develops a specific (tactical, precise) engineering approach, within that basic framework, for evaluating various intercrop designs.

Some questions of terminology

While there is some variability in the use of terms amongst intercrop researchers, a certain uniformity seems to have evolved. The following is taken from Andrews & Kassam (1976), with modifications and commentary as needed for the purposes of the developments in this book. The most general term is multiple cropping, under which is the dichotomous classification of sequential cropping and intercropping. *Multiple cropping*: the general term, is '... growing two or more crops on the same field in a year'. *Sequential cropping*: the time dependent form of multiple cropping, is '... growing two or more crops in sequence on the same field per year ... Crop intensification is only in the time dimension. There is no intercrop competition. Farmers manage only one crop at a time in the same field.' *Intercropping*: the space-dependent form of multiple cropping, is '... growing two or more crops simultaneously on the same field. Crop intensification is in both time and space dimensions. There is intercrop competition during all or part of crop growth. Farmers manage more than one crop at a time in the same field.' Under the general category of intercropping there are four subcategories. (1) *Mixed intercropping*: 'growing two or more crops simultaneously with no distinct row arrangement'. This is frequently the form taken in indigenous slash and burn or fallow agriculture (Rappaport, 1971; Russell, 1968), as depicted in Figure 1.1(*a*). (2) *Row intercropping*: 'growing two or more crops simultaneously where one or more crops are planted in rows'. This is the pattern usually encountered in intensive agriculture, where the plough has replaced the machete and fire as the main tool of land preparation, as depicted in Figure 1.1(*b*). (3) *Strip intercropping*: 'growing two or more crops simultaneously in different strips wide enough to permit independent cultivation but narrow enough for the crops to interact agronomically'. This form of intercropping is more common in highly modernized systems, especially where the intensive use of machinery is desired, as depicted in Figure 1.1(*c*). (4) *Relay intercropping*: 'Growing two or more crops simultaneously during part of the life cycle of each'. This form of intercropping may actually include the other three as subsets, since its primary categorization variable is time.

One could challenge this particular taxonomy of multiple cropping on several bases. It may, for example, make more sense to think of three general categories: (1) sequential cropping, (2) relay intercropping, and (3) full intercropping. Such a taxonomy would be based on the degree of physical

Fig. 1.1. Typical intercropping situations (*a*) a mixed garden of the Mari in highland New Guinea (Rappaport, 1971); (*b*) row intercrops of corn and beans in Chile; (*c*) strip intercrops of corn and wheat in central China.

association of the crops involved, ranging from no association (sequential) to partial association (relay) to complete association (full). But it probably does not make much difference in the long run, and I am not willing to go down in history as yet another scientist who willingly participated in an argument about terminology. As far as this book is concerned we accept the terminology offered by Andrews & Kassam (1976) in which multiple crops are either sequential or intercrops, and intercrops in turn are mixed, row, strips, or relay.

Given the above taxonomy it is useful to clarify to which sorts of intercrops the theory developed in this book applies. As far as the theory has been developed it is most directly applicable to mixed, row, and strip intercropping. The abysmal lack of a dynamic component in most of the theoretical formulations here mitigates against the theory being applied at the level of relay intercropping, although some small steps to rectify this error of omission are presented in Chapters 11 and 12.

In addition to the major categories of intercrops as detailed above, there has also grown an entire lexicon associated with intercropping research. Here I take the approach of being reasonably liberal with the use of various terms, perhaps more than would suit most other workers. My feeling is that cautious use of terminology includes the judicious appreciation of how terms are in fact used in the literature, not a preaching insistence on the use of a particular rigid terminology. Some terms seem well-established and one need not be concerned about confusing one or another reader. Other terms are rigidly adhered to by some segments of the scientific community but eschewed by others (e.g., use of the term 'sole' crop by workers who publish in the journal *Experimental Agriculture* is ubiquitous, yet someone accustomed to working in the United States, especially in the south, might find his or her audience subconsciously envisioning intercrops of grits and greens – the term 'soul' food looms large in our legends). Thus the following list of terms is meant to guide the reader to their use *in this book*, with not the slightest implication that they should or should not be used in this form elsewhere. I have tried to incorporate the most general usage of each term. The terms are:

Monoculture (sole crop) – the cultivation of a single species of crop.

Intercrop (polyculture) – the cultivation of two or more species of crop in such a way that they interact agronomically (biologically). Intercrops can be of four flavors – mixed, row, strip, or relay – as indicated above.

Cropping pattern - the yearly sequence and spatial arrangement of crops or of crops and fallow on a given area.

Cropping system – the cropping patterns used on a farm and their interaction with farm resources, other farm enterprises, and available technology which determine their makeup.

Land equivalent ratio (LER, relative yield total, RYT) – the ratio of the area needed under monoculture to a unit area of intercropping at the same management level to give an equal amount of yield (see chapter 2).

Relative value total (RVT, income equivalent ratio, IER) – the ratio of the area needed under sole cropping to produce the same income as one hectare of intercropping at the same management level (see Chapter 2).

Competition (interference) – the process in which two individual plants or two populations of plants interact such that at least one exerts a negative effect on the other (see below).

Facilitation – the process in which two individual plants or two populations of plants interact in such a way that at least one exerts a positive effect on the other. Double facilitation is equivalent to mutualism.

Of all the above terms, competition perhaps needs a bit more elaboration. One school (we'll call it the American school) thinks of competition as simply the negative effect of an individual or population on another individual or population. Frequently, competition is subdivided into exploitation competition (in which both populations or individuals are attempting to exploit the same or similar resources), and interference competition (in which one or the other population or individuals interferes in some way with the well-being of the other, for example through shading or the production of allelochemicals) (see any elementary ecology text from the United States or Canada). Another school, (what we might call the British school), reserves the word competition for those situations in which the populations or individuals are in competition *for* something (such as a critical resource) (see, e.g., Harper, 1977). This, school uses the word 'interference' as the general term for a negative interaction. Thus what is competition for the Americans is interference for the British, and what is exploitation competition for the Americans is simply 'competition' for the British, and what is interference for the Americans is that part of interference that is not competition for the British. At any rate it is all quite confusing and I do not want to add to the confusion by insisting on one or the other usage, nor proposing my own. In this book, then, both competition and interference are taken to be general terms that refer to a negative interaction between two individuals or populations. The use of either term does not imply some extra knowledge about 'what is being competed for' nor how one is interfering with the other. I shall eschew usage of the subdivisions (either exploitation or interference of the American school or 'competition for something' of the British school). Thus any reference to *either* competition or interference refers to the general phenomenon of negative interaction.

Overview of the general theory

The basic theoretical problem, as treated in this book, is one of organism–environment interaction. It is a double transformation problem in which the organism and the environment affect, or transform one another. A mass moving in a space–time continuum is a good analogy. In an Einsteinian universe the mass responds to local cues in space–time. Yet it has an important effect on that very same space–time. In the same general way, a plant (or any organism for that matter) lives according to the dictates of its local environment, yet is an important participant in effecting change on that local environment.

This simple idea has been presented as a general formulation for plant competition (Goldberg & Werner, 1983) and can clearly be seen in many earlier writings of ecologists, especially those concerned with succession. For example,

> By the term reaction is understood the effect which a plant or a community exerts upon its habitat. . . . It is entirely distinct from the response of the plant or group, ie., its adjustment and adaptation to the habitat. In short, the habitat causes the plant to function and grow, and the plant then reacts upon the habitat, changing one or more of its factors in decisive or appreciable degree. (Clements, 1928, p. 79).

In an excellent summary statement Harper (1977, p. 354) notes:

> A plant may influence its neighbors by changing their environment. The changes may be by addition or subtraction and there is much controversy about which is more important. There may also be indirect effects, not acting through resources or toxins but affecting conditions such as temperature or wind velocity, encouraging or discouraging animals and so affecting predation, trampling, etc.

In Figure 1.2 the idea is presented in a simple schematic form. In Figure 1.1(*a*) the environment–organism transformation is shown for a two-species intercrop situation. The bean, for example, has an effect on the environment (e.g. it may take up potassium, thus leaving an environment partially depleted of potassium, or it may enhance the environment by leaving bits of nitrogen that it had fixed from the air). Both the bean and the corn must respond to this effect, thus setting up the dichotomy of the 'effect' and the 'response' (Goldberg & Werner, 1983).

Both the effect and response have obvious modifiers, as indicated in Figure 1.2(*b*). Many of the engineering applications in this book will be concerned with these modifiers. But the central focus of the abstract theoretical development is not concerned with these modifiers, but, rather, with the qualitative nature of the effect and response. Specifically, this qualitative nature is formulated as a dichotomous classification. First, an organism may affect the environment in a negative way, with respect to other organisms. For example, through nutrient extraction a slightly depleted environment is created, or through the production of shade or allelochemicals a possibly benign environment becomes hostile for another individual. We call this effect competition, recalling the caveats above.

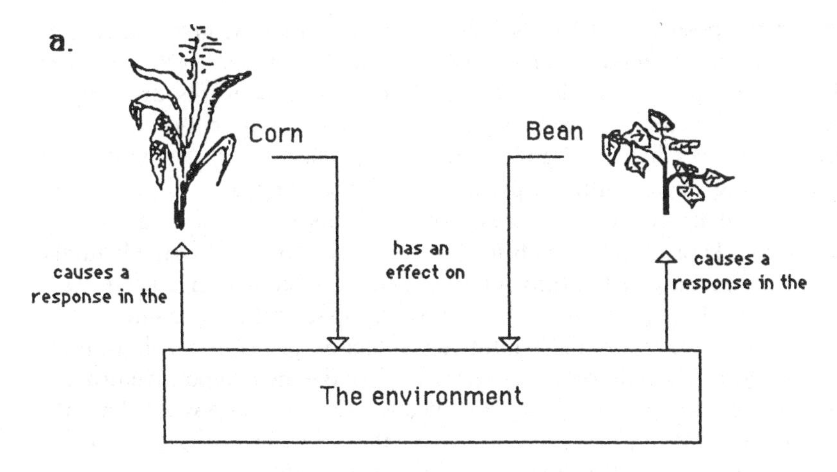

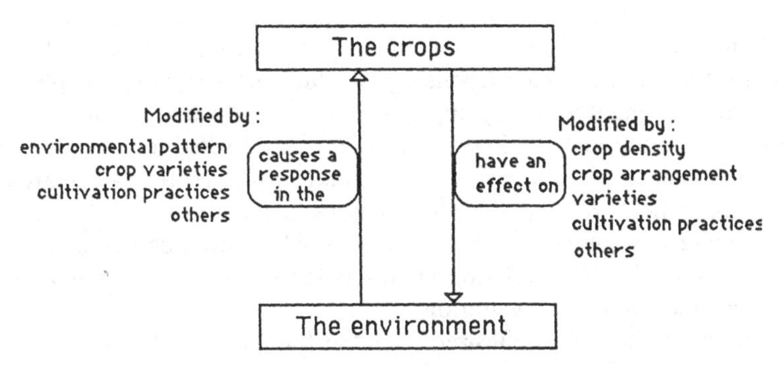

Figure 1.2. Diagrammatic representation of the effect-response formulation.

Second, an organism may affect the environment in a positive way, with respect to other organisms. For example, pollinators attracted to one flowering individual may create a pollinator-filled environment for another (Rathcke, 1984), or the cover provided by paloverde trees protects the saguaro cactus from freezing winters (Vandermeer, 1980*a*). This effect we call facilitation.

The dichotomus theory to be developed is based on a few well-known ideas from theoretical ecology, basically ideas surrounding the general problem of how species interactions might create the structure of communities. Two core ideas dominate: first, when one species has an effect on the environment which causes a negative response in the other species, yet both can more efficiently utilize necessary resources when living together than when in monoculture, we have the *competitive production principle* (or the 'interference' production principle, Vandermeer, 1981*a*); second, when the environment of one species is modified in a positive way by a second species such that the first is facilitated by the second, we have the *facilitative production principle* (Vandermeer, 1984*a*) or, simply, facilitation. It is not that these concepts are new to intercropping research. Indeed, they have been at least implicitly stated many times before (see Chapters 3 and 4). Rather, I propose them as a core to the understanding of how intercrops function, formalizing them so that their quantification in any arbitrary case is obvious. Thus, if the effect and response (see Figure 1.1) result in competition, and the intercrop shows an advantage over the monocultures, that advantage is due to the competitive production principle. But if the effect and response result in facilitation, the intercrop is advantageous because of the facilitative production principle. In short, if the environment is affected negatively we possibly have the competitive production principle operative, whereas if it is affected positively we possibly have the facilitative production principle operative (that they do not operate *necessarily* is discussed in later chapters).

An important generalization that comes from this formalism, which is amplified considerably in later chapters, is that interventions (e.g. intercrop improvement, the design of new intercropping systems, etc.) are likely to have dramatically different outcomes depending on which of these two principles is operative, thus making it arguably important to know which is which. Yet, as discussed fully in Chapter 4, it is in principle impossible to distinguish between the two using the simple plot trials so common in intercropping research. Thus, this theoretical framework, while offering a basis in which the vast literature on intercropping can be newly interpreted, simultaneously suggests a new set of empirical criteria, offering a new set of questions which might open alternative avenues of experimentation.

However, it should be emphasized that the theory offered here is only intended to speak to a subset of the variables that are important in intercropping systems. The various social aspects are virtually ignored, with the trivial exception of simple economic conversion of yields to prices. The important long-term effects of protection from soil-erosion, maintenance of genetic resources, watershed management, and a host of other considerations are, similarly, not treated. The theory is aimed strictly at the agronomic and ecological aspects of intercropping systems, those aspects which, in my opinion, are most easily understood in the context of accepted ecological theory (sometimes appropriately modified ecological theory).

Finally, some comments are in order with regard to the general philosophy of the theory as presented herein. Theoretical formulations must accord with their proposed use. For some purposes what might be described as phenomenological models or black boxes will be most appropriate. For example, when two crops are suspected or known to form combinations that have the potential for producing well in an intercrop because they compete only weakly (i.e., no special facilitative environmental modification is known or suspected), it will be most useful to formulate their relationship in terms of generalized competition coefficients, as is done in Chapter 11. For other purposes, the details of the competition might permit a more rational use of an external input such as nitrogen or water. In such a case theoretical formulations must be more mechanistic, that is, must include the operation of the factor of concern. In general, I have taken the point of view that a range of theoretical formulations from the most phenomenological to the most mechanistic is optimal, making models available at whatever level might be appropriate for particular applications. Thus, Chapters 3 and 4 might appear overly phenomenological to more reductionist thinkers, while Chapters 5 and 6 might appear overly reductionistic to those purporting a holistic philosophy. My position is that any phenomenon is better understood when a battery of nested models is available (Levins, 1968).

Future use of intercrops

There seems to be a prejudice among casual observers and intercropping researchers alike that intercropping is for peasant farming and has no place in modern agriculture. I am violently opposed to this idea. It is not true and perhaps only seems true for the very reasons I have outlined earlier, the lack of a theoretical focus that might guide empiricism.

When 'modern' agriculture involves varieties specifically adapted for production in monoculture, machines specifically adapted for production in monoculture, and research methodology specifically adapted for improvement of monocultures, what might one expect? This subject is treated in more detail in Chapter 12. For now, suffice it to say that until modern production technology is developed, including some sort of theoretical foundation for the agronomic aspects, as this book purports to do, it will be a *fait acompli* that intercropping will have no place in modern agriculture. When research technology for intercrops is as well-developed as it is today for monocultures, and when machines plant and harvest intercrops, and when specific varieties have been developed for their performance in intercrops, intercropping will no longer be just for peasant producers.

Structure of the book

In Chapter 2 we review some of the basic concepts involved in measuring intercrop performance. Some of the concepts which may initially seem somehow overly practical will be seen in later chapters to be the jumping-off point for various theoretical explorations. Chapters 3 and 4 are really the core of the book. In those two chapters I present the two principles on which all of the subsequent theoretical developments depend. Chapters 5 and 6 explore a slightly more mechanistic approach to understanding the two main principles. Chapters 7, 8, and 9 explore some special topics with respect to the two basic principles. Chapter 7 examines those intercropping situations in which perennials are involved, emphasizing the shade-casting properties of overstory perennials. Chapter 8 treats the situation of intercrops and weeds, actually a special case of the competitive production principle in which three components, crop one, crop two, and weeds, are in competition. In Chapter 9 we examine the question of intercrop variability, a topic referred to repeatedly in previous chapters but here developed as an integrated whole. Finally, Chapters 10, 11 and 12 look at the possibilities of using the developed theory in the improvement of existing intercrops and in the possible design of new intercrops.

2

The measurement of intercrop performance

Is there an advantage to growing intercrops? The simplest answer to this question is the qualitative one. If so many traditional agriculturists do it, there must be some advantage to it. This attitude is fine at the most general level, and we really need go no further. But at a more specific level we wish to ask the question about particular intercropping systems. Is it or is it not true on a particular farm in northern Brazil that corn grown together with peanuts is better than corn grown alone and peanuts grown alone? That is the question on which we focus in this chapter.

The bases for comparison

The problem of population density and planting design

Whatever the method of evaluation, the underlying basis is always a comparison of the performance in intercrop to the performance in monoculture. The first complication arises when one must decide what monoculture production figures should be used in the evaluation. In Figure 2.1 several possibilities are illustrated. In Figure 2.1(*a*), the polyculture is stipulated first, and we are to decide which monocultures to use for the computation of the land equivalent to ratio (LER). If we use case I, the overall population is the same in monoculture and intercrop. If we use case II, the overall population density is larger in the intercrop than it is in the monocultures. Consider the similar Figure 2.1(*b*). In this case we begin with the monocultures as stipulated, and are to decide which intercrop to use. A moment's reflection reveals the similarity between the problems as posed in Figure 2.1(*a*) and that posed in Figure 2.1(*b*). Typically, case I (of either Figure 2.1(*a*) or (*b*)) is called a 'substitutive design' and case II an 'additive design'. Note that the question of additive vs substitutive design is based entirely on population density, a point which is problematical and to which we return later. For now we note that stipulating either additive or substitutive solves the problem of what monoculture to use for comparison. In a substitutive design the monoculture

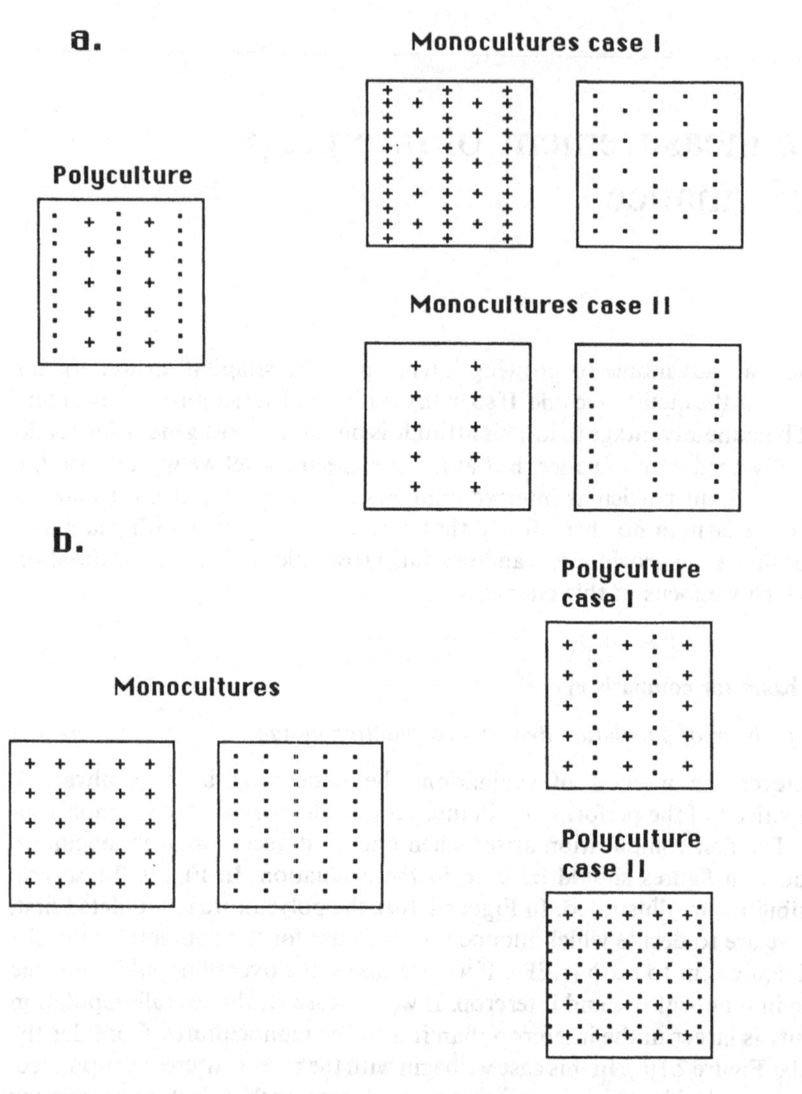

Fig. 2.1. Diagrammatic illustration of additive vs. substitutive designs: (*a*) Given a design for the polyculture, the monocultures can be constructed by replacing all positions in the intercrop with the single species (case I = substitutive design), or simply by eliminating the other species from the polycultural design (case II = additive). (*b*) Given the designs for the two monocultures, the polyculture can be constructed by planting half the positions in the polyculture with one species and half with the other (I = substitutive design), or by planting both species exactly the way they are planted in the polyculture (case II = additive design).

is constructed by simply removing one of the species from the intercrop. In an additive design the monoculture is constructed by substituting the like species for all positions where the other species occurs in the intercrop. We can begin with the intercrop as given, and from it construct the appropriate monocultures (Figure 2.1(*a*)), or begin with the monocultures and from them design the appropriate polycultures (Figure 2.1(*b*)). In either case the stipulation of additive or substitutive is sufficient to define unambiguously what monoculture should be used in the LER computation.

The dichotomy additive vs substitutive is a convenient one and in fact covers much of the experimental work that has been published to date. But from another point of view it is hopelessly restrictive. For example, corn and bean production in southern Mexico is typically done as an intercrop with corn and beans planted in the same hole with the holes approximately one meter apart in a grid (Vandermeer, *et al.*, 1983). If an additive design is used, the monocultures would be one each of corn and beans, both planted as a grid with 1M spacing. If a substitutive design were to be used, each of the monocultures would have two plants (of the same species) at each point on the same grid. But it is likely that in practice, if a farmer were to plant monocultures of these two crops, the population density would be neither the substitutive nor additive design, but, rather, something in between. Imposing an additive design would probably mean that fewer plants than could be profitably grown in the given area would be planted. Imposing a substitutive design would probably mean that more plants than could produce at maximum efficiency would be grown. Thus the farmer would most likely choose monocultures somewhere in between the substitutive and additive design.

The point here is that a whole range of possibilities are available. Whether you begin with a stipulated intercrop and seek the monocultures to which it should be compared, or whether you begin with fixed monocultures and seek some sort of 'good' intercrop design, the additive/substitutive dichotomy is unnecessarily restrictive. But if we discard it, we are again left with the question of which monoculture should be used in computing LER for a particular intercrop (or the parallel question, which intercrop should be planted in the first place).

There is no fixed solution to this problem. It is quite important in intercropping research to indicate the criterion upon which monocultures and/or intercrops are selected, but there is no way to stipulate that any particular design is, in some sense, correct. Nevertheless, some informal and unwritten rules seem to have evolved among intercropping researchers.

For most agronomic or practical applications, it is usually assumed that the monocultures used for comparison are the 'optimal' monocultures, that is, the monocultures that produce the highest yield (Huxley & Maingu, 1978; Mead

& Willey, 1980). Recalling the basic relationship between yield and density in monocultures, that density at which the yield peaks is the density that should be used in the monocultural comparison. On the other hand, for strictly biological interpretations one frequently presumes an additive design (case II of Figure 2.1). Regardless of such unwritten conventions (many researchers would probably deny their existence anyway), there is ultimately no compelling reason to use one or another monoculture – it depends on the ultimate interpretation one wishes to apply. For the remainder of this book, unless stated otherwise, the agronomic assumption of optimal monoculture (i.e. the one with the highest yield) will be assumed.

But while the problem of what monoculture to use can be solved by insisting on the optimal one, it is not easy to stipulate the 'optimal' intercrop. Since there are virtually an infinite number of possible designs, and because the decision criteria of what really is optimal vary so greatly (see below), the problem is not very tractable. By restricting the problem in one way or another, the optimal solution will sometimes be obvious. For example, if one crop is fixed in its design (e.g. because of a restriction on mechanical planting or harvesting), finding the optimal design for the second crop becomes identical with the design of a monoculture. But frequently, such restrictions may lead to an optimization problem that is theoretically tractable but unfeasible from a practical point of view. Such an example is provided in Chapter 10, in which a heavily parameterized model for a tomato–soybean intercrop is theoretically capable of providing the optimal solution, but, because of the heavy parameterization, the utility of the model is probably marginal at best.

The yield set

Instead of searching for, or defining what might be an optimum, or what restrictions might make it possible to find one, it is better, at least conceptually, to represent all possible yield combinations of all possible intercrop designs. That is, suppose we could do thousands of different experiments with thousands of different intercrop designs, theoretically with every design possible. After plotting the resulting yields of all those experiments on a graph of yield vs yield (of each of the two crops), we would obtain a set of points which would represent all possible intercrop combinations, as pictured in Figure 2.2. This set of points is called the *yield set* (Vandermeer *et al.*, 1984) and is similar to the composite production possibility curve so common in economic analyses (e.g. Raintree, 1983) (actually the edge of the yield set is the production possibilities curve).

Viewing the yields in this fashion we see the general possibility of having a continuum, the ends of which are defined by two extreme situations. The yield

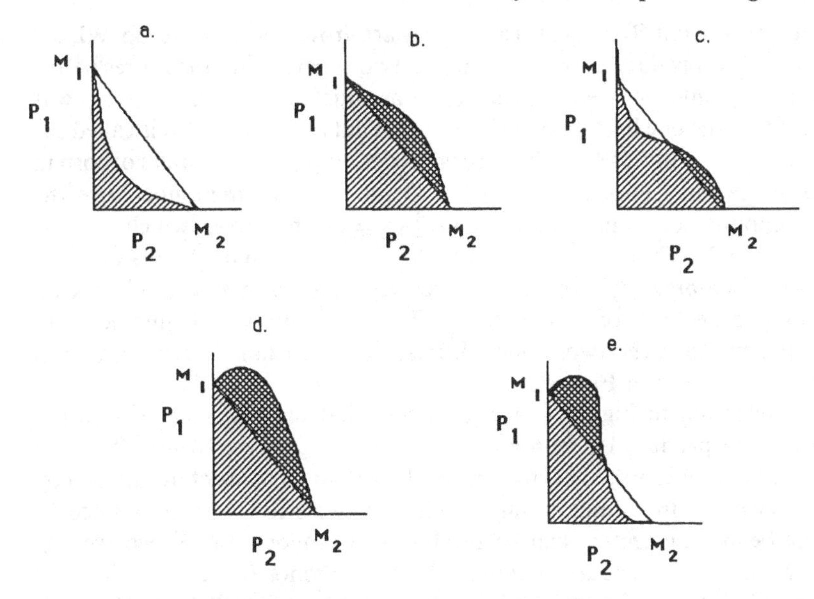

Fig. 2.2. Examples of qualitatively different forms of the yield set. In all cases, cross-hatching indicates combinations that give LER greater than unity.

set might be concave, as in Figure 2.2(*a*) or it might be convex, as in Figure 2.2(*b*). Figure 2.2(*c*) illustrates an intermediate situation of a yield set that is partially convex and partially concave. A qualitatively different sort of yield set is obtained when facilitation is involved in the system. Figure 2.2(*d*) presents a convex yield set in which the second crop facilitates the first crop, while in Figure 2.2(*c*) we have the same facilitative interaction but with a concave yield set. The heavily shaded areas representing yield combinations above the line from M_1 to M_2, represent all possible cases of intercrop advantage under the criteria of the land equivalent ratio (LER), as discussed in the next section.

Criteria for intercrop advantage

The land equivalent ratio (LER)

The measurement most frequently used to judge the effectiveness of an intercrop is the land equivalent ratio (LER) (Mead & Willey, 1980). It takes its name from its interpretation as relative land requirements for intercrops versus monocultures (as described below, the relative yield total, RYT, is identical). Let us suppose that on one hectare of land it is possible to produce

10 units of corn and 50 units of beans if they are grown as an intercrop. What if one wanted to produce corn and beans as two separate monocultures? How much land would be needed to produce as much in monocultures as was produced on the one hectare of polyculture? That amount of land is called the land equivalent ratio. If $\frac{3}{4}$ of a hectare is needed to produce 10 units of corn in monoculture and $\frac{1}{2}$ a hectare to produce 50 units of beans in monoculture, the total amount of land needed is $\frac{3}{4}+\frac{1}{2}=1\frac{1}{4}$ hectares which means $LER = 1\frac{1}{4} = 1.25$. That is, to obtain the 10 units of corn and 50 units of beans that could be obtained from a single hectare of intercrop, one would need a total of 1.25 hectares of monocultures. The intercrop would thus be more advantageous than the two monocultures. Another example is presented in diagrammatic form in Figure 2.3.

With reference to Figure 2.3 it is apparent that since 20 units of corn per hectare can be produced in monoculture, in order to produce 10 units (to equal a unit area of intercrop), $\frac{1}{2}$ hectare is needed, or 10 units per hectare, divided by 20 units per hectare. Similarly (again with reference to Figure 2.1), since 75 units of beans per hectare can be produced in monoculture, $\frac{2}{3}$ hectare ($\frac{50}{75}$) would be needed to produce 50 units of beans in monoculture. Thus the total number of hectares of corn and bean monocultures needed to produce the equivalent of a single hectare of corn–bean intercrop is $\frac{1}{2}+\frac{2}{3}=1.17$, which is the land equivalent ratio.

From the above example it is easy to see that the land equivalent ratio is simply the sum of the relative yields. That is, the relative yield of corn (RY_{corn}) is its yield per acre in intercrop (P_{corn} in Figure 2.3) divided by its yield per acre in monoculture (M_{corn} in Figure 2.3), or $RY_{corn} = P_{corn}/M_{corn}$. Similarly $RY_{bean} = P_{bean}/M_{bean}$, and, finally,

$$LER = RY_{corn} + RY_{bean} = P_{corn}/M_{corn} + P_{bean}/M_{bean}. \qquad (2.1)$$

The land equivalent ratio (or relative yield total) is thus easy to compute and easy to interpret. If it is greater than 1.0, the intercrop is more efficient. If it is less than 1.0, monocultural production is more efficient. The value 1.0 is the critical value, above which the intercrop is favored, and below which the monocultures are favored.

The problem can be understood most generally by making reference to the yield set. Since the yield set stipulates all possible combinations of the crops in question, it naturally must include the 'best' one. The best one is also defined as (at least here for the moment we take its definition to be) the maximum possible value of the LER. We can graphically interpret this maximization procedure by referring to equation (2.1), rewriting it as

$$P_1 = (M_1) - (M_1/M_2)P_2,$$

which is a linear equation in P_1, P_2 space (the subscripts 1 and 2 are generalizations of the subscripts corn and bean of equation (2.1). Since we wish

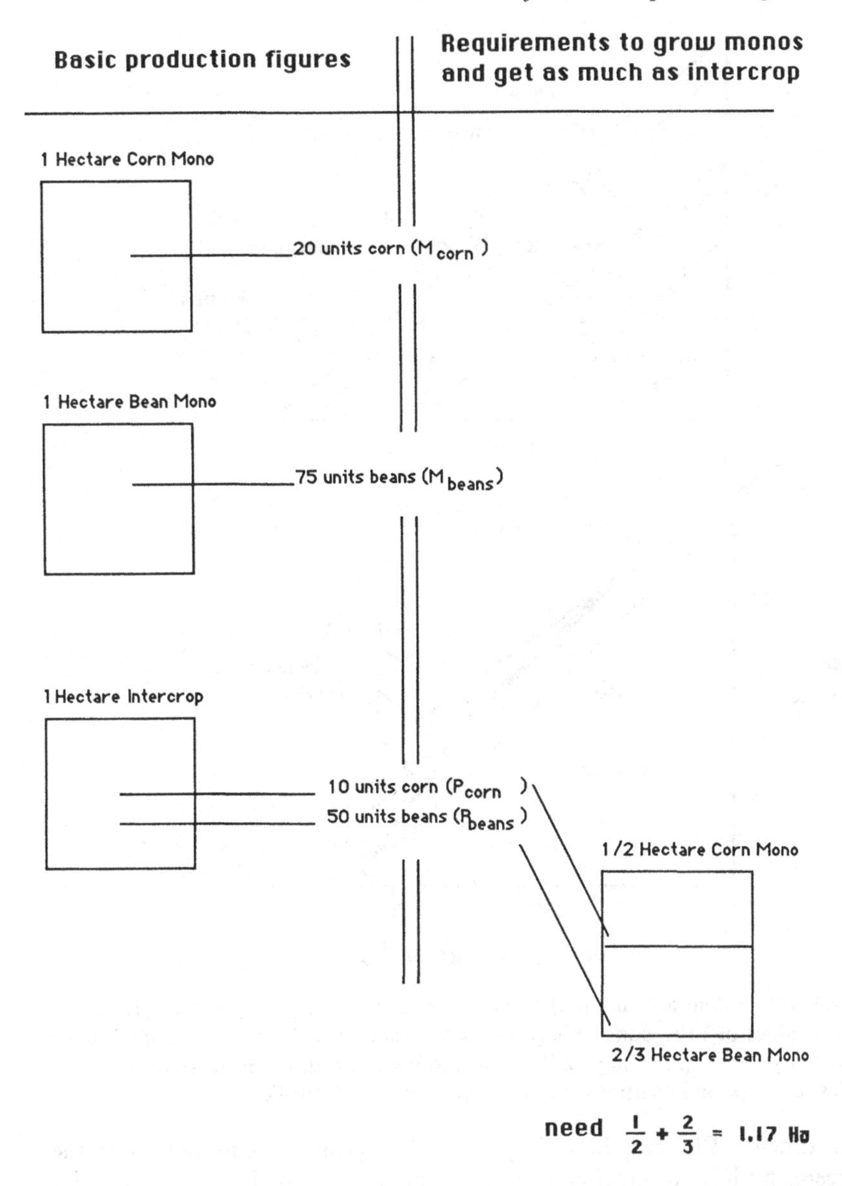

Fig. 2.3. Illustration of the meaning of the land-equivalent ratio (LER).

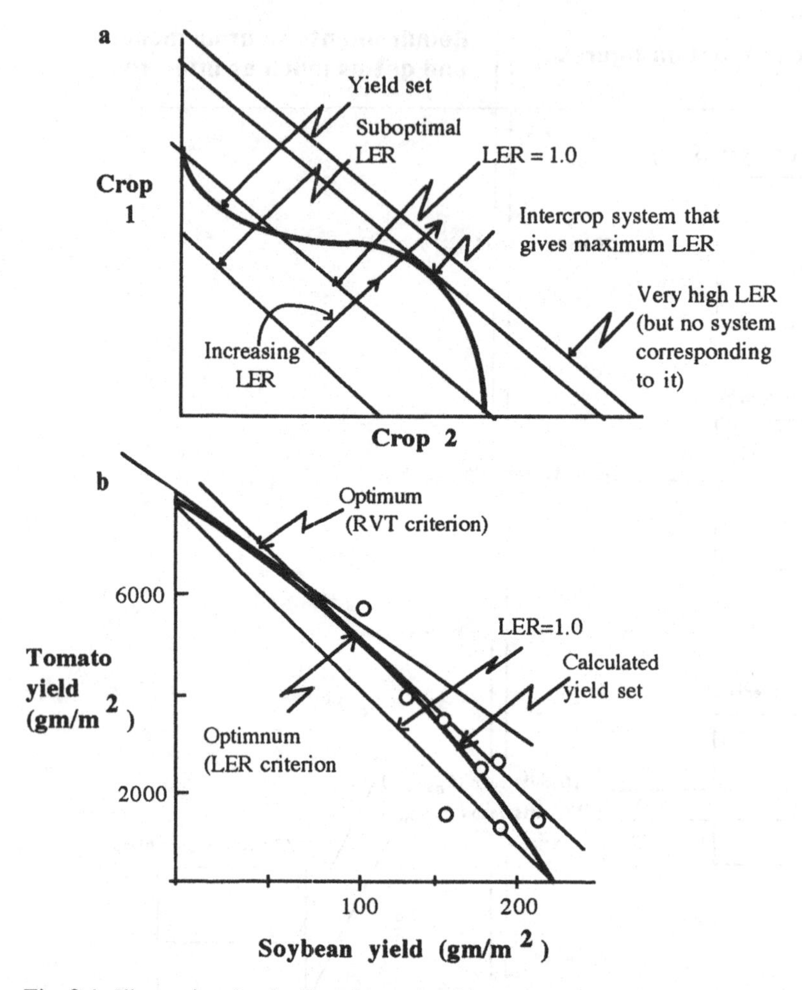

Fig. 2.4. Illustration for finding the optimal intercrop using the yield set (*a*) various values of LER, some suboptimal, one supraoptimal, and one optimal; (*b*) a calculated yield set (see Chapter 10 for methodology) with experimental values (circles) and optimal solutions (from Vandermeer *et. al.* 1984).

to maximize LER and the only place LER fits in the equation is as the intercept, finding its maximum is the same as finding the largest possible intercept. This process is illustrated for a general case in Figure 2.4(*a*), in which a maximum is determined on a concave–convex yield set. It is possible to conceive of a value of LER so high that no real possibilities exist for its realization, as in the case of the top line in Figure 2.4(*a*). It is also possible to conceive of a value of LER that is low enough to be realizable with the two crops in question, but lower than other possible realizable values, as in the case of the bottom line in Figure 2.4(*a*) (the 'suboptimal' LER). It is exactly that line that is tangent to the yield set that gives the optimal (largest possible) LER, as labeled in Figure 2.4(*a*). Figure 2.4(*b*) shows a calculated yield for a tomato–soybean combination (calculated from theoretical principles described in Chapter 10), along with the position of the optimal intercrop.

Other criteria of success

Further complications are introduced when one allows that, indeed, different criteria are sometimes required. One common situation is when the producer is concerned about only one of the two crops. In other words, the yield of that crop in the intercrop must be larger than its yield in the monoculture, or $P_{corn} > M_{corn}$. For example, it is common in India for a farmer to require a *full* yield of cereal crop (i.e. that which would be obtained in monoculture) and to plant a legume intercrop only if the yield of the cereal will not suffer (Mead & Willey, 1980). Naturally, any situation in which a second crop would increase the yield of the principal crop (i.e. $RY_{principle\ crop} > 1.0$) would be desirable under this criterion. For example, in The Gambia (West Africa) peanuts are intercropped with sorghum because a weevil that attacks the sorghum is apparently diverted by the confusing effect of the peanuts (Steedman, pers. comm.) (see Chapter 4). Without the peanuts the sorghum is apparently unable to produce at all, which means that RY approaches infinity (in a sense, any obligate symbiotic association has an RY approaching infinity).

A second situation frequently not fully appreciated is when the producer requires particular mixtures of produce, either because of market conditions, labor management, or dietary requirements (Riley, 1985; Willey, 1985; Berkowitz, pers. comm.). For example, pasture improvement through the incorporation of a legume into grass-dominated pasture carries with it a practical restriction. Because of nutrient rquirements of cattle, the final yields must contain between 25 and 35% legume in the forage (no greater amount). Thus, the criterion whereby we decide if an intercrop is good, is restricted by another factor. The restriction is, formally, $P_1 < a_1(P_1 + P_2)$ and $P_1 > a_2(P_1 + P_2)$, where a_1 and a_2 represent the limits on the desired proportions. Thus we have two linear functions

$$P_1 < [a_1/(1-a_1)]P_2,$$
$$P_2 > [a_2/(1-a_2)]P_2,$$

which define an area of acceptability on a graph of P_1 vs P_2, as indicated in Figure 2.5. If a_1 and a_2 are equal, we simply have a straight line, and the optimum is where that straight line intercepts the yield set.

Perhaps the most common situation in the contemporary world has to do

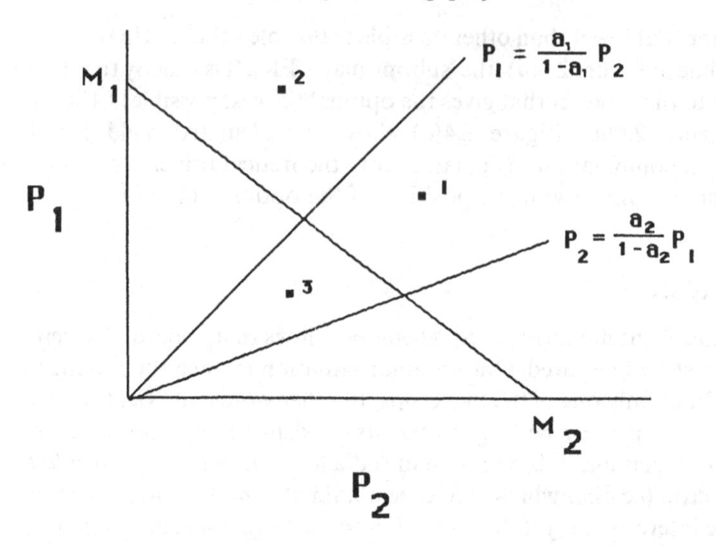

Fig. 2.5. Illustration of restriction placed on intercrop yields by external requirements, as stipulated by the two equations (see text). Point labeled 1 represents intercrop advantage and acceptable yield combination. Point 2 represents intercrop advantage and unacceptable yield combinations. Point 3 represents acceptable yield combinations but no intercrop advantage.

with monetary values, not biological yields. Since the two terms in the LER are relative, an LER computed from the crop's monetary value is in fact identical to the LER computed from biological yields. But, more importantly, when a producer is concerned about monetary value, the intercrop should be compared to the most valuable of the two monocultures. In the example from Figure 2.1, if corn is valued at 10 pesos per bushel and beans at 1 peso per pound, the producer could make 200 pesos per acre from a corn monoculture (20 bushels × 10 pesos) and only 150 pesos per acre with an intercrop (10 bushels × 10 pesos + 50 lbs × 1 peso = 100 + 50 = 150), despite the fact that the LER is greater than 1.0. An index that takes this goal into account is the 'relative value total' (RVT) (Schultz *et al.*, 1982; also known as the income equivalent ratio, IER, Andrews & Kassam, 1976), and is computed as follows:

$$RVT = (aP_1 + bP_2)/aM_1,$$

where a is the price of the first crop, b is the price of the second crop, and $aM_1 > bM_2$. The critical value of RVT is 1.0. If it is greater than one, the intercrop is advantageous, if it is less than 1.0, the monoculture has the advantage. Note that if LER is less than 1.0, RVT *cannot* be greater than 1.0. LER is thus sort of a base-line measurement, and is usually computed before

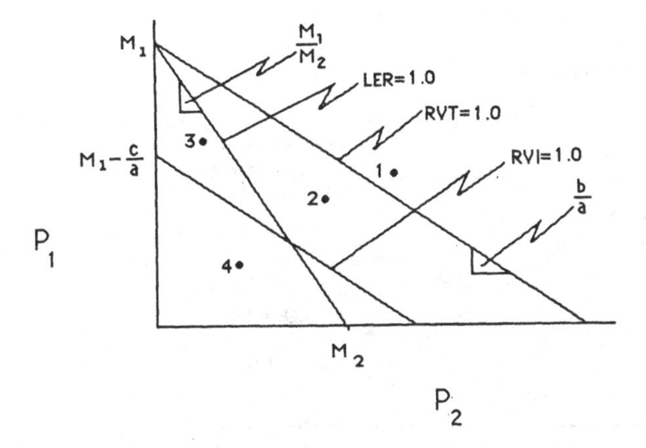

Fig. 2.6. A comparison of three criteria for intercrop advantage, LER, RVT, RVI (see text for definitions). The point labeled 1 represents an intercrop system which is advantageous under all three criteria. Point 2 satisfies the RVI and LER criteria, point 3 only the RVI criterion, and point 4 satisfies none.

anything else. If LER is small (less than 1.0) there is no need to compute RVT for it must also be less than 1.0. In general, RVT is less than LER, under all circumstances.

Particularly in terms of modern applications of intercrop technology, it is frequently desirable to use an intercrop to replace some other agricultural input. Thus, for example, when corn is planted as a trap to induce *Heliothis* out of cotton (Pearson, 1958), the interest is in whether the net production in the intercrop will be better than the net production in the monoculture, where we must factor in the cost of insecticide. Or, if a secondary crop is used to reduce or eliminate herbicides in a primary crop, likewise the net polycultural production must be compared to the monocultural production minus the cost of the herbicides.

Thus, we may define a 'replacement value of the intercrop' (RVI) as

$$RVI = (aP_1 + bP_2)/(aM_1 - c),$$

where c is the cost of the input that the intercrop is designed to replace. $RVI > 1.0$ is the criterion for intercrop advantage. Unlike the relative value total, the replacement value is not limited to those situations in which $LER > 1.0$, as is made clear in Figure 2.6.

These various considerations of measuring intercrop yields can be conveniently summarized on a single graph, as I have done in Figure 2.7.

All of the above implicitly assumes that a producer is simply interested in maximizing yield. Another apparent function of intercropping is to avoid risk.

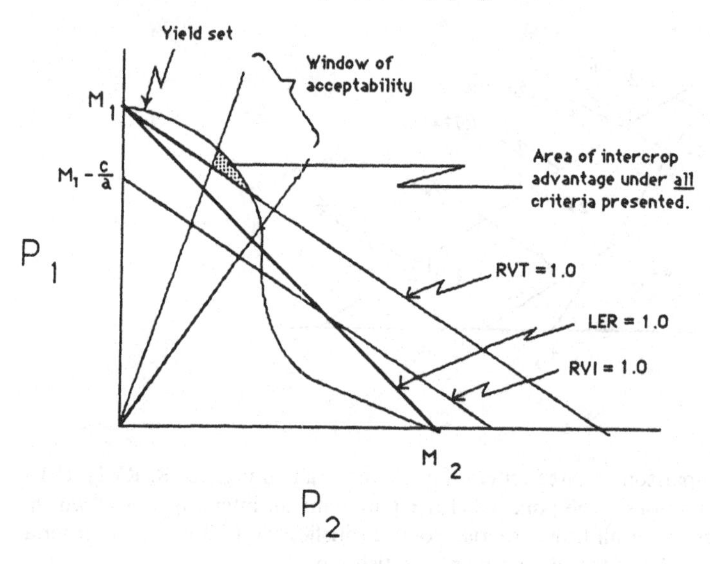

Fig. 2.7. The various criteria of intercrop evaluation.

That is, a producer might be willing to sacrifice a somewhat larger monocultural yield if he or she can gain some security. If corn produces better in monoculture than in polyculture, but approximately every four or five years a disease unpredictably wipes out all of the corn, the farmer who has intercropped with beans has at least something, while the moncoculturalist has nothing. This question is much more difficult to deal with than the question of simply maximizing yields, and is taken up as a problem in measurement and theory in Chapter 9.

Some recurrent statistical problems

It has been pointed out repeatedly that statistical problems are encountered when dealing with measures of the sort described above (Oyejola & Mead, 1982; Pearce & Gulliver, 1978; Fisher, 1977; Mead & Stern, 1981). To demonstrate some of these problems consider the data presented in Table 2.1. These data are purposefully unrealistic so as to demonstrate numerically some of the statistical problems, and show how they might be quite severe. (The data are actually taken from a lognormal distribution in such a way that the expected value of LER is 1.0). The data are meant to mimic an experiment in which there are five replicates, each of which contains a monoculture and a polyculture. Computing LERs for each of these replicates we obtain the last column of Table 2.1(*a*). All values of LER are greater than 1.0 and the mean is

Table 2.1. *Hypothetical example of LER computations, illustrating some statistical problems (see text)*

a. Crop I Monoculture	polyculture	*RY*	Crop II Monoculture	polyculture	*RY*	*LER (RYT)*
403	55	0.14	1	3	3.00	3.14
1339	1	0.00	4	818	204.50	204.50
148	999	6.75	5431	12	0.00	6.75
99	70	0.70	3754	1220	0.32	1.02
30	2	0.07	4	246	61.50	62.20
						$\bar{X}=55.52$

$t=2.573$ (after a log transformation), $p=0.032$, Mann–Whitney $U=0$, $p=0.005$.

b. Crop I Monoculture	polyculture	*RY*	Crop II Monoculture	polyculture	*RY*	*LER (RYT)*
404	55	0.14	1838	3	0.00	0.14
404	1	0.00	1838	818	0.44	0.44
404	999	2.47	1838	12	0.00	2.47
404	70	0.17	1838	1220	0.66	0.83
404	2	0.00	1838	246	0.13	0.13
						$\bar{X}=0.80$

$t=-1.479$ (after log transformation), $p=0.175$, Mann–Whitney $U=5.00$, $p=0.059$.

55.52. This is a ridiculously high value of LER, all the better to see the problem. Remember, with these data, there in fact is no yield advantage to the intercrop and the LER should be 1.0. After a log transformation, a *t*-test shows the mean LER to be significantly greater than 1.0, at least at the 0.05 level. With such data an experimenter might very well conclude that this is a situation of intercrop advantage, and a simple *t*-test of the observed LERs against the critical value of 1.0 verifies that conclusion. But in this case the 55.52% intercrop advantage is in fact a statistical artifact.

If one examines the data in Table 2.1 closely, the origin of this statistical problem becomes transparent. Since the LER is a sum of ratios, if a very large number appears by chance in the numerator, the ratio will be extraordinarily large. But if that same number appears in the denominator, the ratio will be somewhere between 1.0 and 0, effectively constrained to be greater than zero.

For example, suppose we have a population of 100 yields, 50 of which are 1.0 and 50 of which are 29, making the mean of the population 15. Let us suppose that it is just chance variation that causes the 14-point deviation from the mean value, either $15-14$ or $15+14$. We might reasonably expect that the average of the ratio of two randomly drawn yields is 1.0, since their mean is always expected to be 15 and the only variation is a chance variation of plus or minus 14. But in a series of samples of pairs of values, if we form the ratio between the two, about 25% of the time we will get two 1s, about 25% of the time two 29s, and about 50% of the time one 1 and one 29. Thus we expect, with 25% chance, each the following ratios, 1/1, 29/29, 1/29, and 29/1. Thus the mean that we actually would find would be $0.25\ (1.0)+0.25\ (1.0)\ (0.034)+0.25\ (29.0)=7.76$, rather than the expected 1.0. This is the reason for the large LERs in the final column of Table 2.1*a*. In fact, there is no yield advantage to the intercrop (that was the way the random sampling scheme was set up), and it is only because of this 'ratio' problem that an experimenter might be fooled into thinking there was. Standard statistical tests do not help much, as is evident in this example.

One procedure that has been used (e.g. Schultz *et al.*, 1982) is to compute relative yields from the mean monoculture yields. The monoculture yield is thus treated as a parameter, so that it does not enter the equation as a variable. This would appear to eliminate the ratio problem since each relative yield is divided by a constant (the mean of the monoculture yields) and thus will not be a ratio of variables. Applying this procedure to the data of Table 2.1*a*, we obtain the figures in Table 2.1*b*. Obviously this procedure eliminates the problem of artificially high values of LER. But, unfortunately, it involves a different problem, perhaps not so severe, but a problem nevertheless. Note that the mean value of LER is not 1.0 as we expect, but, rather, 0.80. Indeed this procedure will always underestimate the true value of LER, since any excessively high yield values, while no longer contributing to very high relative yields, contribute to the high (relatively speaking) value of the denominator, which is then applied to each ratio. Clearly, this problem is not as severe as the other, (a *t*-test, for example, fails to detect a significant difference between this 0.80 value and the expected 1.0), but it is a real problem nevertheless. Investigators using mean monoculture yields in computing relative yield totals should thus regard their estimates as conservative (i.e. on the low side).

There are many other statistical problems associated with the analysis of intercrops. Analyzing all of them would take an entire volume, which is exactly what has been prepared by the statistician Federer (1987) whose excellent volume should be consulted by those interested in further elaborations. The particular problem of bias in the LER is presented here because of the central position of the LER criterion in developing the theory presented in the next chapter.

3

The competitive production principle

The basic idea

We begin our exploration of the mechanisms of intercrop advantage by pointing out that one potential mechanism is, in a sense, no mechanism at all. While casual observation leads us to question what causes particular patterns, it is sometimes worthwhile to reflect and ask if the patterns perhaps occur because nothing much is happening. It turns out, as described below, that because of the way we define intercrop advantage, it is quite possible, even perhaps common, to observe an intercrop advantage without anything special happening.

It is easiest to understand this 'mechanism' by referring to a similar phenomenon, recognized for many years, in community ecology. It is generally accepted (although the details remain hotly debated) that when two species do similar things (i.e. occupy the same niche, interfere with each other's activities, compete with one another, etc.) it is unlikely that there is enough room in the environment for both. Loosely, two species cannot occupy the same niche. If their niche requirements are sufficiently similar, which is to say they compete with one another intensely, one or the other will become extinct, given a long enough time. On the other hånd – and this point is often not emphasized sufficiently – if the two species have similar but distinct requirements, which is to say they compete with one another only weakly, they may both persist indefinitely in the environment. That they cannot both persist when competing intensely has been referred to as the competitive exclusion principle, although it is obvious that the emphasis could have been placed on their coexistence, suggesting a 'competitive togetherness' principle.

The competitive exclusion principle is frequently represented graphically as a two-dimensional graph of the biomass (or yield or population density) of one species plotted against the biomass of the other species. Such a graph is presented in Figure 3.1. On each axis we first plot the carrying capacity of each of the species, that is, the biomass that each is expected to reach in the environment when the other is absent. So for the value of 0 on the X axis (i.e. when there is a zero biomass of the first species) we plot the value of the

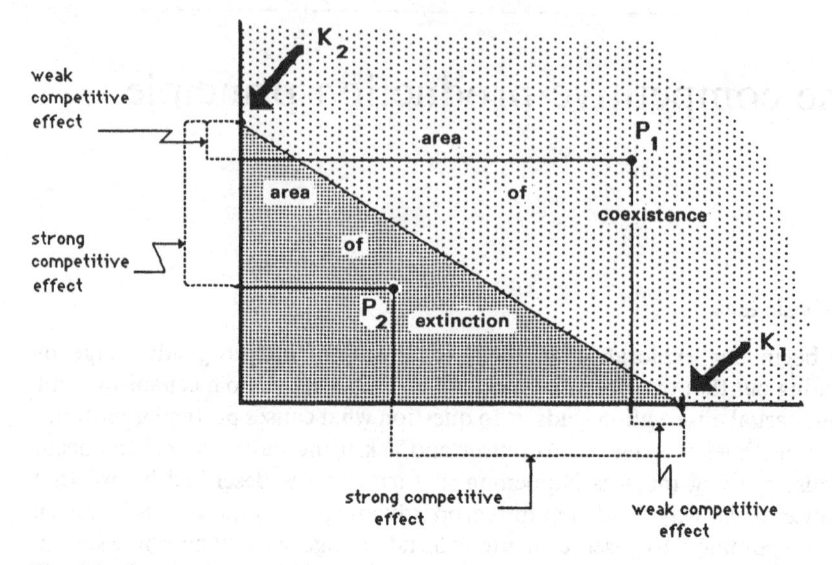

Fig. 3.1. Standard graphical evaluation of species coexistence–extinction in a 2-species competitive situation. All points falling above the line connecting the two carrying capacities represent combinations that will coexist. All points falling below that line represent combinations for which one or the other species will go extinct. The axes represent the population densities (or biomasses) of each of the populations.

carrying capacity of the second species (usually symbolized by K) on the Y axis. Similarly when there is no biomass of the second species (the value on the Y axis is 0), we plot the K (carrying capacity) of the first species on the X axis. The process of interspecific competition can be seen as a reduction from the carrying capacities (i.e. what happens to the biomass when it is affected by competition from the other population). Thus what we expect is a reduced value along each axis, representing the biomass of the affected population. In this way the biomass of each species in competition is represented by a point on the graph. If competition is intense, we expect that the encumbered biomass will be much lower than the unencumbered biomass (carrying capacity). If competition is weak, we expect that the encumbered biomass will be only slightly lower than the unencumbered biomass. Both of these situations are illustrated in Figure 3.1. The interesting feature of this method of presentation is that the representation of the encumbered biomass on a two-dimensional graph provides an instant assessment of the underlying principle of either coexistence or extinction. If the point representation of the encumbered biomasses falls above a line connecting the two unencumbered biomasses (which will generally happen when competition is weak), the two species will likely coexist. If the point falls below the line (which will generally happen when competition is strong), one of the two species must be driven to extinction. This entire process is illustrated diagramatically in Figure 3.1.

A great deal of ecological research has focused on themes related to the competitive exclusion principle, not necessarily by name, but certainly in principle. It was originally formulated by the Russian physiologist Gause (1934), later popularized (Hardin, 1960), and extensively studied experimentally (e.g. Colwell & Fuentes, 1975). The exact details are variable, and certainly the simple abstract approach of Figure 3.1 has seen an explosion of complicating and enriching features (e.g., MacArthur, 1972; Levins, 1968; Vandermeer, 1970; Case & Gillpin, 1974; May, 1974). Its utility has been challenged and its very existence seriously questioned on both theoretical and empirical grounds. But, nevertheless, much of modern ecology takes as its point of departure the veracity of the general proposition that if competition is sufficiently weak, the two species may coexist, whereas if competition is sufficiently strong one or another of the species will perish – the competitive exclusion principle.

It is difficult to ignore a similar principle with regard to intercrops. Recalling the basic definition and critical values of the land equivalent ratio, as discussed in chapter 2, the critical value of LER is 1.0. If it is above 1.0 the intercrop is said to be advantageous, if it is below 1.0 the monoculture holds the advantage. Using the defining equation of LER, we can write the critical value

$$LER = (P_1/M_1) + (P_2/M_2) = 1.0,$$

where P_1 is production of species 1 in the intercrop, P_2 that of species 2, M_1 production of species 1 in monoculture, and M_2 that of species 2. Multiplying both sides by M_2 and rearranging, we obtain

$$P_2 = M_2 - (M_2/M_1)\,P_1. \qquad (3.1)$$

Equation (3.1) is linear in the two variables P_1 and P_2. Furthermore, the intercepts on both axes are the unencumbererd yields (the monocultures), and the graph of equation (3.1) itself is a straight line between the two monocultures. The criterion of LER greater than unity, as established earlier, can thus be interpreted graphically quite easily. If the point representing the intercrop yields of the two crops is above the line connecting the unencumbered yields, the intercrop is said to overyield ($LER > 1.0$). If the point representing the intercrop yields of the two crops is below the line connecting the unencumbered yields, the intercrop is said to underyield ($LER < 1.0$). This criterion is illustrated graphically in Figure 3.2. As can be seen from that diagram, a biological explanation for intercrop advantage could simply be the same as the biological explanation for species coexistence – if the competitive

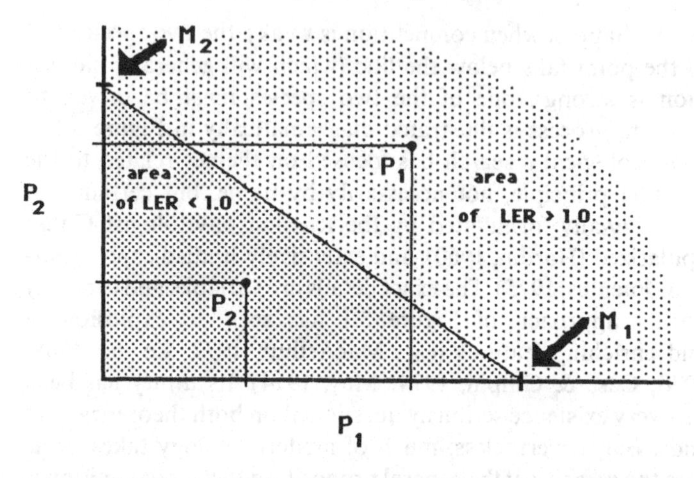

Fig. 3.2. Standard graphical evaluation of yield advantage in an intercropping system. All points falling above the line connecting the two monocultural yields (M_1 and M_2) represent combinations that overyield. All points falling below that line represent combinations for which intercrops are inferior to their monocultural alternatives.

pressure is sufficiently weak the intercrop will be advantageous. Such a biological rationale is identical to the rationale for the competitive exclusion principle, thus giving rise to the term 'competitive production principle' (Vandermeer, 1981*a*), and also referred to herein simply as the mechanism of reduced competition. The principle is diagrammed so as to emphasize its similarity to the competitive exclusion principle in Figure 3.2.

In the ecological literature the 'weakness' or 'strength' of competition is invariably cast in terms of the relative intensity of interspecific competition as compared to intraspecific competition. The competition coefficients of the classic Lotka–Volterra equations are in fact ratios of inter to intraspecific competition (e.g. Vandermeer, 1981). In the context of intercropping it might be useful to recall that this ratio is what is important. On the other hand, given that monocultures are almost always treated as if they were at optimum yields (see Chapter 2), and that intercrop yields are compared with their monocultural counterparts, the question of intraspecific competition is largely cancelled out of the picture.

With the mechanism of reduced competition in mind we see how an intercrop that shows an advantage over its monocultural components according to the common LER criterion may not require any special explanatory mechanism, any more than the coexistence of two species requires any further explanation other than that they do not affect one another strongly enough to cause extinction. Yet, in the sense that they are grown in the same field but do not compete intensely, that lack of competition may require some sort of explanation. Thus, one might search for a mechanism of differential resource allocation or niche differentiation, a topic explored further in Chapter 5. The competitive production principle can be thought of as a guide to practical design (or perhaps *is* such a guide amongst peasant producers). More importantly, it suggests ways in which ecological theory can be brought to bear quantitatively on the practical problem of intercrop design, creating a sort of 'ecological engineering', one of the subjects of Chapter 10.

Some interesting antecedents

The same idea, or something very similar, had been elucidated by Trenbath (1976) citing much earlier literature (Ludwig, 1950). Trenbath called the phenomenon 'annidation' and defined it as '. . . that where intercrop components exploit the environmental supplies of growth factors in differing ways'. As long as 'growth factor' is defined broadly enough, annidation and reduced competition are equivalent. Fisher (1975) suggested that 'this (intercrop advantage) can only be true if those environmental resources which limit yield are available in greater quantity to the mixed crops than to the pure stands'. While not absolutely correct, this statement attributes all cases of intercrop overyielding to reduced competition (see also Loomis, 1984). Similar statements, at least approximately equivalent, have been common in various reviews of intercropping (e.g. Kass, 1978; Willey, 1979*a*, *b*; Trenbath, 1974).

It is very likely that many intercrop systems are designed, albeit informally, with this principle in mind. For example, in a general description of mixed cropping in the tropics, Norman *et al.* (1984) state:

> The exploitation of micro-environments . . . by crops with specific requirements or tolerances is ingenious. Thus shade-tolerant species are planted on the shady margin of the plot, moisture-demanding species at the bottom of sloping sites, fertility-demanding species on localized ash concentrations, . . . and climbing species against unfelled trees or beside rigid upright crops.

In personal conversations with peasant producers in southern Mexico, Costa Rica, and Nicaragua, I have frequently been struck by what seems to be an informal use of the competitive production principle. I have been told that two crops make a good combination because one is taller than the other and 'fits in' to the spaces where the other does not, or that the root systems go to different depths and thus use nutrients from different parts of the soil. Such popular knowledge is apparently quite common among peasant producers and is really quite the same idea as reduced competition.

Niche theory, effect–response and yield sets

A great deal of ecological research has been concerned with the relationship between competition and ecological niches (Vandermeer, 1972; Colwell & Fuentes, 1975; Pianka, 1976). The idea is really quite simple. If the niches of two species are too similar, they probably compete sufficiently intensely to exclude one another, or, in the immortal words of Dr Seus,

> And Nuh is the letter I use to spell Nutches
> Who live in small caves, known as Niches, for hutches.
> These Nutches have troubles, the biggest of which is
> The fact there are many more Nutches than Nitches.
> Each Nutch in a Nitch knows that some other Nutch
> Would like to move into his Nitch very much.
> So each Nutch in a Nitch has to watch that small Nitch
> Or Nutches who haven't got Nitches will snitch.

It thus becomes interesting to focus on ecological niches as a vehicle for understanding competition (which in turn dictates the pattern of intercrop advantage).

Much of the modern theory concerning niches ultimately derives from the remarkable insights of G.E. Hutchinson (1957). Hutchinson conceived of a niche as a multidimensional space, the coordinates of which are those environmental factors that are important to the species of concern. Most importantly, Hutchinson recognized the central role of biological interactions by distinguishing between the niche prior to interaction and the niche after interaction, the *fundamental* versus the *realized* niche. The fundamental niche is thus a measurement of a species' performance in an environmental space, prior to its interaction with the other species of concern. The realized niche is that same measure after the interaction. Given these ideas we immediately see how niches and competition are necessarily related.

But the intimate connection between ecological niches and competition is not all that simple. At one extreme we can clearly declare that if two species do not overlap in their fundamental niches they cannot compete. Such a statement is virtually a tautology. But because zero overlap indicates zero competition, it does not then follow that much overlap indicates much competition (Colwell & Fuentes, 1975; Pianka, 1976).

This general problem is illustrated in Figure 3.3. Considering just the extreme situation, there are only four cases of interest. In Figure 3.3(*a*) and (*b*) the overlap in fundmental niches of the two species is great but only in Figure 3.3(*a*) is competition significant. In Figure 3.3(*c*) and (*d*) competition is small due to a small overlap in fundamental niches, but even so, in Figure 3.3(*d*) competition is intense within that subset of the environment that falls within the zone of fundamental niche overlap.

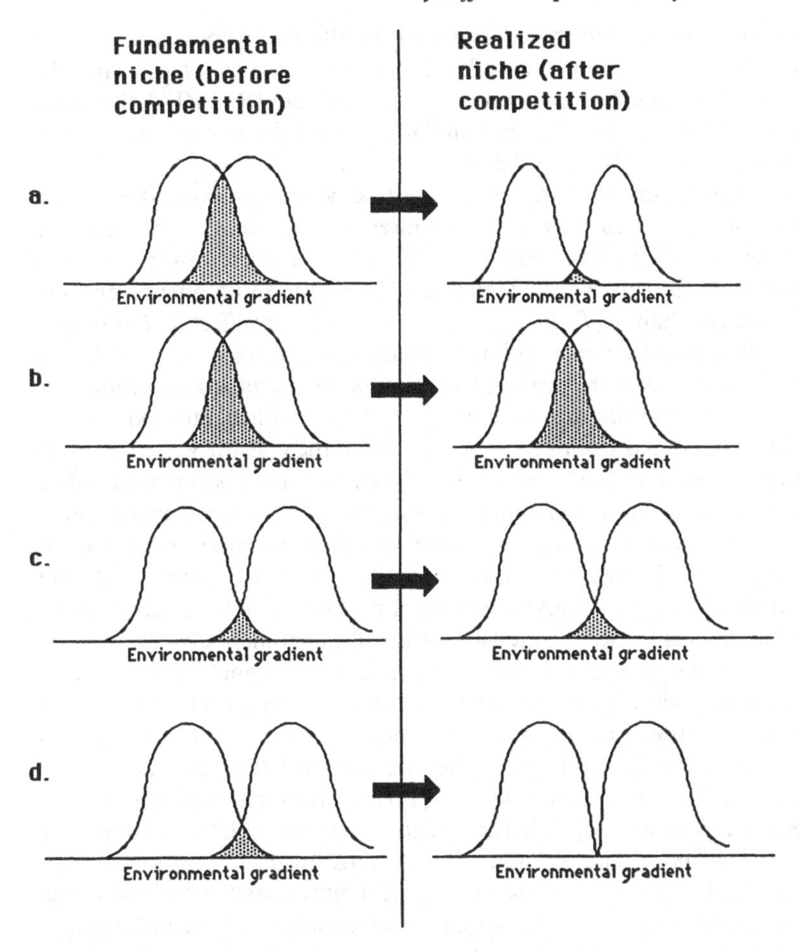

Fig. 3.3. Graphical representation of elementary concepts in niche theory. In each graph the ordinate represents some measures of fitness (e.g. yield) and the abscissa an environmental gradient.

Generally speaking we would expect intercrop advantage to be highest in Figure 3.3(*b*, *c* and *d*). A small overlap in fundamental niches may thus indicate good performance for the intercrop, but a large overlap does not necessarily signify anything, and overlaps in realized niches are simply irrelevant.

An alternative, but related, method of representing these same concepts is in terms of effect and response, as introduced in Chapter 1. Plants interacting with one another can be thought of as engaging in a two-stage process – exerting an effect on the environment and responding to the environment thus affected (see Figure 1.1). If species *A* exerts an effect on the environment, species B must respond to the environment as altered by species A. We can thus think of the environment as a function of species A ($E=f_1(A)$) and the performance species B as a function of the environment ($B=f_2(E)$), the yields (or biomass, or density, or any other similar measure) give rise to environment E which affects the yield of species B.

For some purposes it makes most sense to develop models that explicitly recognize this chain reaction, as in Chapter 5 (and others) when specific mechanisms are elaborated. For other purposes it makes most sense to develop models recognizing the higher level interaction as a phenomenon in itself. Thus in the chain A–E–B, we recognize that A has an effect on B (through E), and that it might be appropriate to study that phenomenon directly. In functional notation we are dealing with simple functional composition. We begin with the crop species B living in the unmodified environment E, ($B=f_1(E)$) unmodified from the point of view of the effect it will experience from crop species A. In the presence of A the environment becomes modified ($E^*=f_2(A)$). We can view that same crop B living in the modified environment E^*, ($B=f_3(E^*)$). Since E is simply $f_2(A)$, we can substitute into f_3 and represent B as $B=f_3(f_2(A))$. We symbolize the composition of the two functions f_3 and f_2 as g, so we can write $B=g(A)$ as the effect of A on B, *indirectly* through its effect on the environment. As noted above, it will sometimes be useful to study the two functions f_2 and f_3 separately (explicitly recognizing the effect – response chain), while for other purposes it will be more useful to examine directly the composed function g (recognizing the higher level interaction as a phenomenon in itself). Both approaches are used in later chapters.

Recognizing the indirect effect of one species on the other through an effect on the environment, as has just been presented, is directly related to the theory of the ecological niche. The function f_1 (i.e. $B=f_1(E)$) is the fundamental niche, while the function g (i.e. $B=f_3\ (f_2\ (A))=g\ (A)$) represents the realized niche.

Furthermore, both the effect–response representation and the niche theory representation can be directly related to the yield set (Chapter 2). Conceive of the problem as the yield (or some other measure) of species A being fixed at some particular value. That value stipulates a certain amount of competition and thus a resultant value of B. Alternatively, we might think of fixing a certain value of B, thus stipulating a resultant value for A. By performing this thought experiment of fixing various values of A and determining the resultant values of B, we are effectively tracing out the g function, described above, that relates the indirect effect of A on B. We could perform the same thought experiment for the effect of B on A, thus stipulating the g function that relates the indirect effect of B on A. This process is illustrated in Figure 3.4 for theoretical populations of corn and beans.

The problem with this conceptualization (and the reason that it can only be a thought experiment) is that if we fix beans at some particular position, we

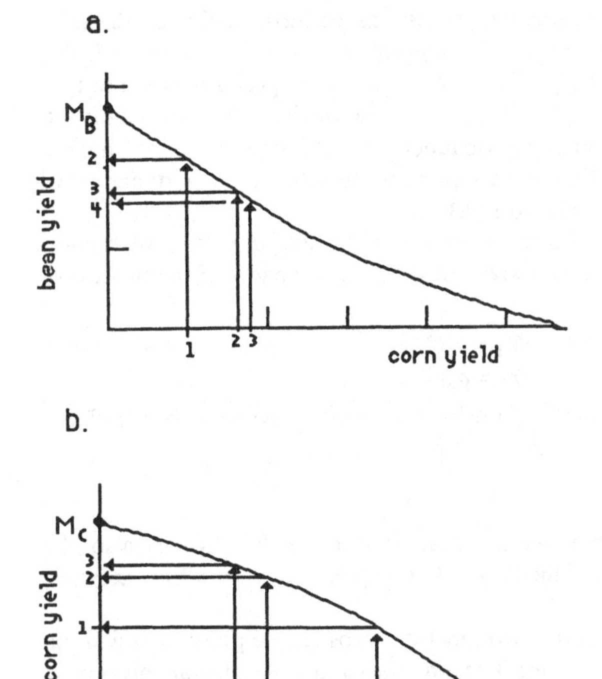

Fig. 3.4. The competition functions relating effect of one species on the other. In both cases the species on the abscissa represents the effector species and the species on the ordinate the receptor species. If the bean yield is set at position 1 on the abscissa of the lower graph, position 1 on the ordinate (corn yield) will be obtained. But if we plot this corn yield on the abscissa of the upper graph we obtain position 2 on the ordinate (bean yield). Position 2 on the ordinate of the upper graph is not the same as position 1 on the abscissa of the lower graph. Plotting position 2 on the abscissa of the lower graph we obtain position 2 on its ordinate, which, when plotted on the abscissa of the upper graph gives position 3, etc.

simultaneously stipulate a value for corn, as illustrated by point 1 in Figure 3.4*b*. But if that point is stipulated for corn, in the other function it may in turn stipulate a value for beans which is not the original value, as illustrated by point 2 in Figure 3.4*a*. Thus, a chain reaction is initiated in which it is generally not possible to arbitrarily stipulate the input. Rather it is an interactive process in which the input of one is determined by the input of the other, and so forth.

The resolution of this interactive process can be most easily visualized by combining the two functions on a single graph, as is done in Figure 3.5. The only place that one can fix a yield and have the other species respond to that yield and initiate its reciprocal effect so as to generate the yield as originally set, is at the point of intersection of the two functions. (This does not imply that the trajectories traced out in Figure 3.5 represent the actual growth dynamics of the two plants, they definitely do not.)

The entire idea can be compactly summarized by noting that the interaction functions, the *g*s, as described above, must balance. That is, if the process of competition is

$$Y_1 = g_1(Y_2),$$
$$Y_2 = g_2(Y_1),$$

the expected yields (or biomasses or other) of the two species can be stipulated as those values of Y_1 and Y_2 that satisfy

$$g_1(Y_2) = g_2^{-1}(Y_2),$$

assuming that g_2 has an inverse. Recall that the g function is also the representation of the realized niche and the composition of the effect–response functions (see above).

Thus, the point at which the two functions cross is the point at which the biomasses of both species balance. The functions g_1 and g_2 will vary according to a host of conditions. For example, if the sowing density of one species is drastically cut, at least the unencumbered biomass ($M = g(0)$) of that species will have to change, if not the entire shape of g_1. If we consider all theoretically possible planting densities and arrangements, we generate a whole set of *g*s, each pair of which will stipulate a single (or possibly several) point. That resultant set of points is the yield set (assuming, of course, that the measure in question is the yield).

At this point a simple numerical example will be useful to clarify these concepts. Let us suppose that a simple linear model can relate the yields of the two crops to one another, which is to say that the competition function g is linear. Thus we have

$$\begin{aligned} y &= M_y - ax \\ x &= M_x - by, \end{aligned} \qquad (3.2)$$

where y is the yield of the first crop, x is the yield of the second, the *M*s refer to the respective monocultural yields, and the a and b are the competition coefficients.

At differing planting densities and designs we expect different values of a and b. (Likewise, at different points along an environmental gradient we expect different values of a and b.) When the density of the first crop (y) is sparse or its pattern is evenly spaced with respect to the other species, we

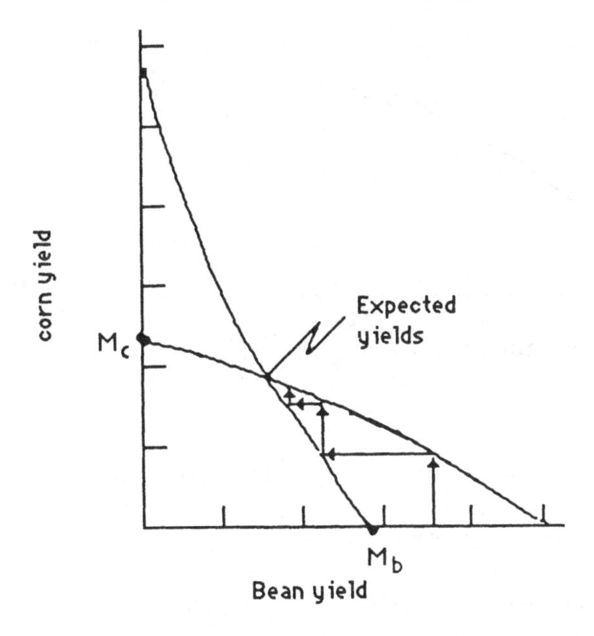

Fig. 3.5. The combined plot of the two competition functions. The only point at which the input and output values correspond is where the two functions intersect.

expect b to be relatively small. When it is dense or its pattern clumped with respect to the other species, we expect b to be relatively large. Figure 3.6(a) gives an example. Similarly, the expected monoculture yields will increase to a point, with increasing population density, as shown in Figure 3.6(b).

Solving equations (3.2) for x and y we have

$$y = (M_y - aM_x)/(1 - ab)$$
$$x = (M_x - bM_y)/(1 - ab), \qquad (3.3)$$

which directly stipulate the yield set. We only have to substitute the various values of a and b for each combination of population densities and solve for y and x.

For example, if y is at a relatively high population density and x is at a medium density (as shown by the two arrows in Figure 3.6(a) and (b)) the value of b will be 1.5, the value of a will be 0.4, the two monocultural yields will be $M_x = 2$, $M_y = 1$. Substituting these values into equations (3.3) we obtain, $y = 0.5$ and $x = 1.25$ (i.e. where the arrows are pointing in Figure 3.6(b)), which is one point in the yield set. By repeating this calculation for all possible combinations of population densities (and thus all combinations of the competition coefficients), we obtain the yield set. The yield set for this numerical example is shown in Figure 3.6(c).

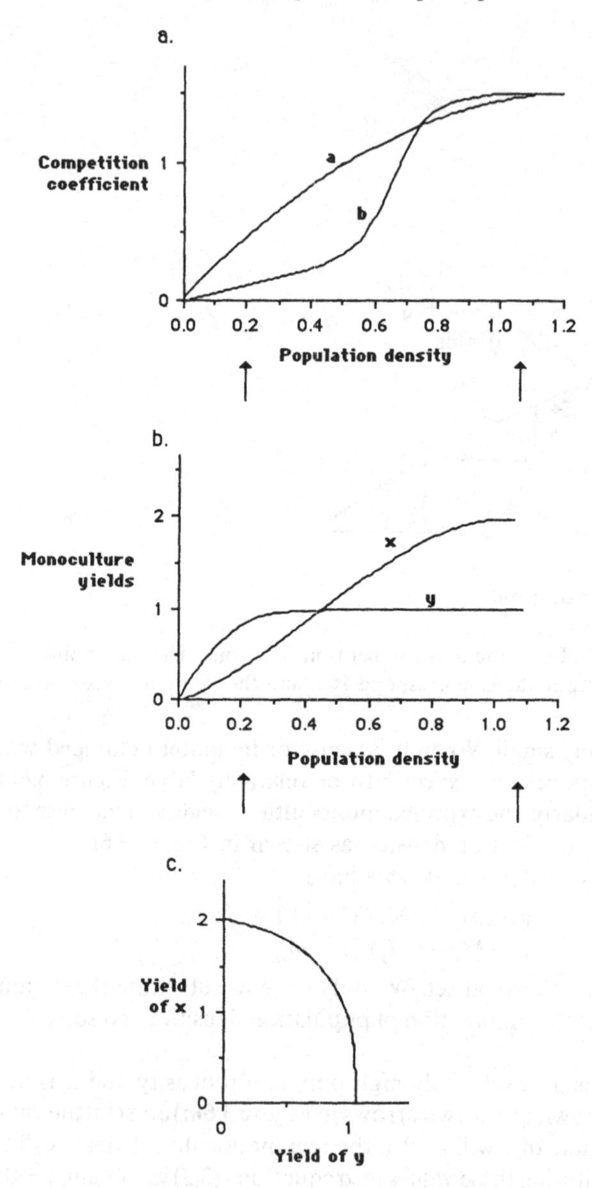

Fig. 3.6. Illustration of the construction of the yield set when the critical environmental factor is population density (*a*) the effect of population density on the competition coefficients; (*b*) the effect of population density on the monoculture yields; (*c*) the resultant yield set (from Vandermeer 1981a).

Up to this point nothing has been said about dynamic theory. The competition functions elaborated here bear a striking resemblance to isoclines of differential equations. In Chapter 12 we demonstrate how, under certain specified conditions, this is true. But it is by no means always true, nor is it a necessary component of the framework developed here. In this section we have simply shown how the elementary processes of effect and response lead to functions which stipulate, ultimately, the yield set, an agronomic concept, and the realized niche, an ecological concept.

Competitive production and environmental grain

Here I introduce a concept that will be central in later chapters, the pattern of the environment, or what is generally referred to as the environmental grain. In a sense the idea is a very old one, simply that environmental factors may or may not occur in clumps or patches, and that when they are patchy they may have an effect on the process of competition. More recently, Levins (1968) and MacArthur & Levins (1967) formalized the concept and invented the term 'environmental grain'.

In this conceptual scheme, environments (or environmental factors or environmental components) are viewed as alternatives. The general idea is not restricted to any particular number of alternatives, but discussion is significantly eased if we allow for only two alternatives. Consider the two alternatives of a very wet versus a very dry patch. If, from the point of view of the organisms of concern, the patches occur as such, that is, if a tree (for example) experiences *either* a wet or dry patch, the grain is called 'coarse'. If, on the other hand, the patches are so small in extent that an individual tree has its roots positioned in some wet spots and some dry spots, from the point of view of an individual tree, the environment is a mixture of patches, and we call it 'fine' grained.

The concept of environmental grain is meant to be quite general, applying over both space and time, and referring to any environmental factor. A useful metaphor for understanding the concept is the moving chessboard, which, when moved back and forth slowly, appears to the eye as the black and white squares that it really is, a coarse-grained pattern. But when moved rapidly, the black and white squares blend, and the eye sees only a shade of grey, a fine-grained pattern.

The grain of the environment is bound to have an important influence on the process of competition, a point explored in considerable detail when we examine the mechanisms of the competitive production principle (Chapter 5).

Applying the principle in nature

Consider, for example, a concrete situation of an intercropping system of maize and cowpea, experimentally studied in southern Mexico (Vandermeer *et al.*, 1983). In that study, three different sites using three different land preparation techniques were employed. Different results were obtained at each of the sites. The three sites were characterized based on the type of land preparation utilized, forest fallow, bush fallow, and grass fallow, depending on how long the particular site had been abandoned. The relative yields and the relative yield totals are shown in Table 3.1.

In the bush fallow site a clear advantage accrued to the intercrop, the grass fallow site showed a clear disadvantage for the intercrop, while in the forest fallow site neither an advantage nor disadvantage was detectable. But what is more interesting are the data on relative yields. We see that in the grass fallow situation the maize was reduced to 45% of its unencumbered yield (i.e. the relative yield is 0.45), while the cowpea was reduced to 41% of its monocultural potential. This means, loosely speaking, that the cowpea exerted a net competitive pressure on the maize which caused a 55% reduction in its potential, while the maize exerted a pressure sufficient to cause a 59% reduction in the potential of cowpea. Such competitive pressure is sufficient to cause underyielding, or to cause the competitive production principle not to operate, indeed suggesting that separate monocultures would be better than the intercrop.

In the forest fallow situation we see a different pattern. The maize was able to reduce the yield of the cowpea to 43% of its monocultural value while the cowpea reduced the maize to 70%, considerably less competition than in the case of the grass fallow. According to the LER criterion in this case the intercrop is neither better nor worse than the monocultural alternatives ($LER = 1.126$, not significantly different from 1.0). A plot of the yields of these two crops in this case would fall exactly on the line connecting the monocultural expectations. In terms of reduced competition, the intensity of competition was not sufficiently small to allow that principle to operate, but also was not sufficiently large to cause underyielding. It was a 'neutral' case.

Finally, in the case of the bush fallow we see a most interesting pattern, one that requires us to think in terms quite different from simply reduced competition. The maize reduced the cowpea to only 20% of its monocultural value, about twice the competitive pressure of the other two situations. But the cowpea not only failed to reduce the yield of the maize, there was an apparent increase in the production of maize as a consequence of its association with cowpea which is to say there was facilitation of the maize by the cowpea. This is a situation that does not fall within the framework of the competitive production principle. The overyielding that occurred was not simply due to the weakness of competition (although that may have been a reality), but was a clear consequence of a facilitative effect exerted on the maize by the cowpea. Whenever a relative yield (of partial LER) is greater than 1.0, something more than competitive production must be operating, which is the subject of Chapter 4.

Table 3.1. *Relative yields and relative yield totals for a maize–cowpea intercrop in southern Mexico* (Vandermeer *et al.*, 1983)

Forest fallow			Bush fallow			Grass fallow		
RYT								
RY_{maize}	RY_{cowpea}	*RYT*	RY_{maize}	RY_{cowpea}	*RYT*	RY_{maize}	RY_{cowpea}	*RYT*
0.35	0.08	0.43	1.23	0.09	1.32	0.21	0.22	0.43
0.51	0.35	0.85	1.45	0.10	1.55	0.18	0.32	0.50
0.59	0.45	1.04	1.91	0.10	2.01	0.14	0.40	0.54
0.75	0.61	1.36	1.09	0.16	1.25	0.70	0.38	1.08
0.78	0.66	1.44	1.54	0.17	1.71	0.42	0.60	1.02
0.95	0.26	1.21	1.30	0.25	1.55	1.04	0.56	1.60
0.97	0.58	1.55	2.23	0.30	2.53			
			1.60	0.41	2.01			
Means								
0.70	0.43	1.13	1.54	0.20	1.74	0.45	0.41	0.86

An additional example is afforded by the system of tomatoes and soybeans (Vandermeer *et al.*, 1984), described more fully in Chapter 10. Using small-scale competition experiments it was possible to gain estimates of various parameters of competition. In particular this case was modeled by a specific interaction function, g, as described above. The environmental factors considered to be important were the densities of the two species, and the two interaction functions were easily decomposable into their response–effect form. That is, the yield of each species was well-represented by an equation that involved the densities of the two species. Formally,

$$y = g(x) = K_x - a(D_y, D_x)x$$
$$x = g(y) = K_x - b(D_x, D_y)y,$$

where the Ds are the respective population densities of the two species and represent the 'environmental factors' in the context of this chapter. The yield set then is given by the solution of this simultaneous set of equations, that is, all combinations of y and x. Substituting the actual functions a and b, the predictive equations are given as

$$y = K_y - a(D_y, D_x) = \frac{K_y - K_x A D_x^{a} - K_x B D_x^{b})}{1 + C D_y^{c} + E D_x^{e} - C E D_y^{c} D_x^{e} - A B D_x^{a} D_y^{b}}$$

where A, B, C, E, a, b, c, and e are parameters relating to the process of competition (see Vandermeer *et al.*, 1984, for details). With this somewhat cumbersome equation, and its partner for the other crop (i.e. $x = K_x - b(D_x, D_y)$), it is possible to compute pairs of values for x and y and thus compute the

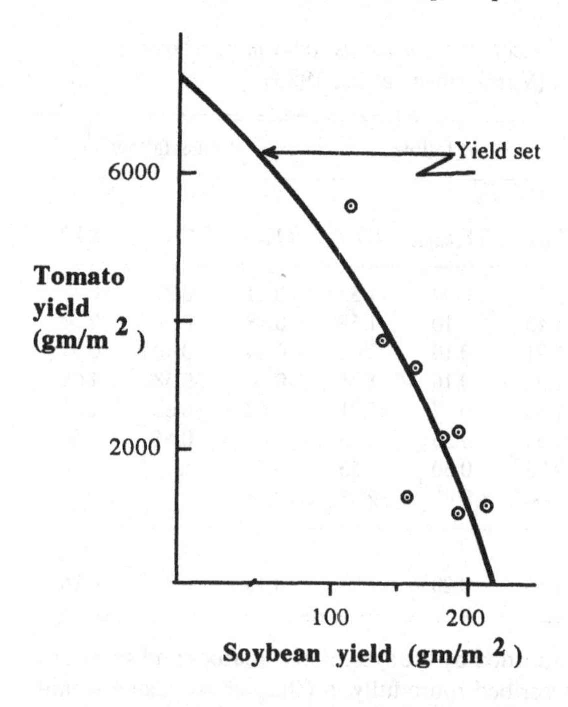

Fig. 3.7. Yield set (for all possible density combinations) as calculated from competition functions, and empirical points from plot experiments (Vandermeer *et al.*, 1984).

yield set analytically. Such a combination is shown in Figure 3.7, along with the observed values for plot experiments. In Chapter 12 I argue that the analytical computation of yield sets (much like this one was computed) could be an important technique for the evaluation of improved varieties for intercrops.

Summary and conclusions

The ideas expressed in this chapter are not surprising. Under certain conditions, a monoculture cannot utilize all the niche space available, and a second crop type can fit in without disturbing the first crop too much. In this way, two crops growing together are similar to two species coexisting in a nonagricultural ecosystem, and Gause's axiom has an obvious counterpart in the agroecosystem. In the same sense that two species will coexist if their mutual competition is sufficiently weak, two crop types will similarly overyield if their mutual competition is sufficiently weak (or, more formally, the interspecific competition is weaker than the intraspecific competition).

If competitive production is operative, some ecological principles could be brought into play both in analyzing particular production patterns and in suggesting programs for increasing production, as discussed in more detail in later chapters. For example, the idea implies that the ecological background can be as important a determinant of overyielding as the crop itself. Thus it makes little sense to speak of particular crop types as overyielding or not, and the mixed results from maize and beans or sorghum and pigeon pea are not at all surprising. Characteristics of the environment in which production is taking place are at least as important as the crop types involved, as one can clearly see in the maize–cowpea example discussed above.

As will become clear in a later chapter, the competitive production principle offers some powerful tools for predicting performance in polycultural systems. Just as Gause's principle can be extended to determine how many species will coexist in an area, the same principle might be extended to find the best crop combinations when large numbers of crops are available. Again depending on the ecological background, the intensity and pattern of competition could be estimated and predictions made as to overyielding potential for various crop combinations. Or, probably more importantly, just as the equations which led to Gause's principle have been expanded and enriched, so might the simple ideas presented in this chapter be expanded and enriched, leading to a greater ability to predict and thus engineer intercropping systems. Such is the idea of several of the later chapters in this book.

4

Facilitation

Figure 4.1 shows a peculiar intercropping system in central Nicaragua. The trees are jocotes (a fresh fruit, *Spondias purpurea*) and the epiphytic cactus growing on their trunks is pitahaya (*Hylocereus* sp.) whose fruit is used to make a drink. At first glance we might think of this as a trivial case of the competitive production principle since neither species would seem to have much of an effect on the other and competition should be quite low.

But a more important, if obvious, observation is to be made here. The pitahaya is an epiphyte and does not grow well on the ground. Thus, the jocote brings something to the environment which benefits the pitahaya – its trunk. The jocote 'modifies' the environment in a positive way for the other species, and thus presents us with a simple and obvious case of facilitation.

The basic idea

This principle is presented as complementary to the competitive production principle, to account for the many cases known both in agricultural (e.g. Vandermeer *et al.*, 1983; Singh *et al.*, 1986; Nair *et al.*, 1979; Osman & Osman, 1982) and in non-agricultural systems (Vandermeer, 1980*b*: Vandermeer *et al.*, 1985; Rathcke, 1984; Hazlett, 1983; Room, 1972) in which one species provides some sort of benefit for another species. Probably there will be particular mechanisms which will not amenably fit into either category, but for the most part it is not difficult to put particular mechanisms in one or the other of these two categories. On the one hand, the two crops use different components of the ecosystem, or use the same components in different ways or in some way exploit distinct ecological niches. This general mechanism, following its analogy to the ecologists' competitive exclusion principle, has been dubbed the competitive production principle (or simply reduced competition). On the other hand, the crops may alter the environments of one another positively (not necessarily reciprocally), and this we call the facilitative production principle (or simply facilitation) (Vandermeer, 1984*b*).

While it is true that an intercrop advantage involving relative yields of less

Fig. 4.1. An intercrop of jocote and pitahaya outside Managua, Nicaragua.

than 1.0 does not require any special mechanism, as discussed in the previous chapter, it is also true that a variety of special mechanisms may nevertheless be operative. It is a question of the difference between necessary and sufficient. It is not necessary to invoke anything further than the competitive production principle if the relative yields do not exceed 1.0. But it also may not be ultimately sufficient to invoke only that principle. Two crops may compete with one another intensely, but only through the protection of one crop from a critical herbivore and the provisioning to the other crop of nitrogen (for example), do they form an advantageous intercrop. These facilitative effects could be strong but the competitive effect sufficiently intense to offset them, leading to relative yields that are less than 1.0. Standard plot trials in such a case would not suggest the operation of anything other than the competitive production principle, even though two other mechanisms would in fact be involved. It is thus critical to develop a conceptual framework that takes full cognizance of the importance of both the competitive and facilitative components of the interaction.

Effect–response and the yield set

Casting the idea in the effect–response mode of previous chapters, if one (or both) of the species in an intercrop causes an effect on the environment to

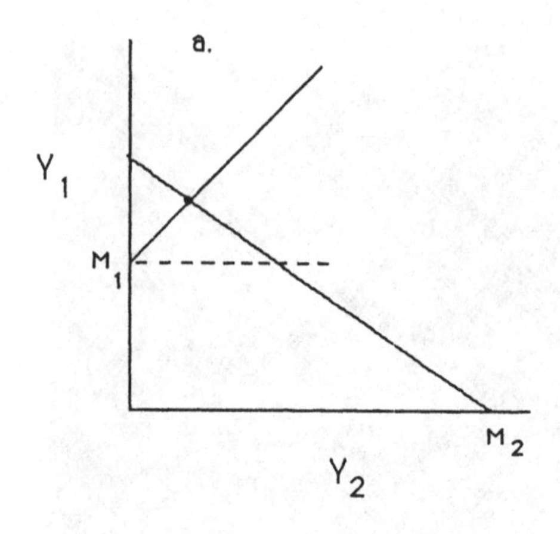

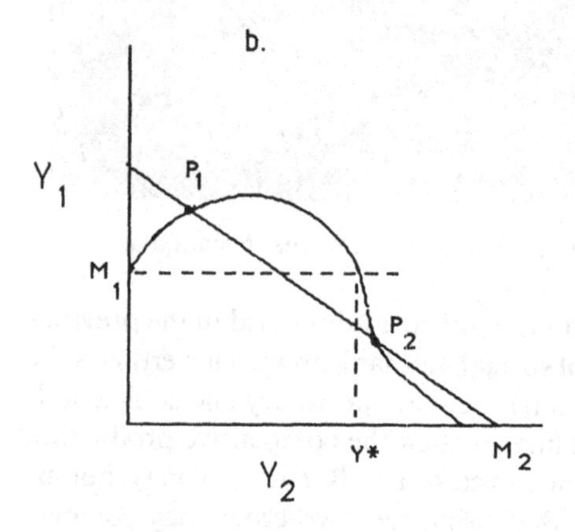

Fig. 4.2. Interaction functions in the case of facilitation. (*a*) Species 2 has a net facilitative effect on species 1 while species 1 has a net competitive effect on species 2. (*b*) At some levels of species 2 (below y^*) the net effect of species 2 on species 1 is facilitative, while at other levels (above y^*) the net effect is competitive.

which the other species responds positively, facilitation occurs. Using the notation developed before, if species *A* exerts an effect on the environment, species *B* must respond to the environment as altered by species *A*. Thus we have the chain *A*–*E*–*B* which represents an indirect effect of *A* on *B*, in this case a positive one. So the 'competition' function *g*, as described in the previous chapter ($B = g(A)$), becomes a 'facilitation' function, and has a positive slope for at least part of its existence.

Figure 4.2 illustrates several examples of the functions g_1 and g_2 (recall from the previous chapter that the biomass or yield of one species is expressed as a

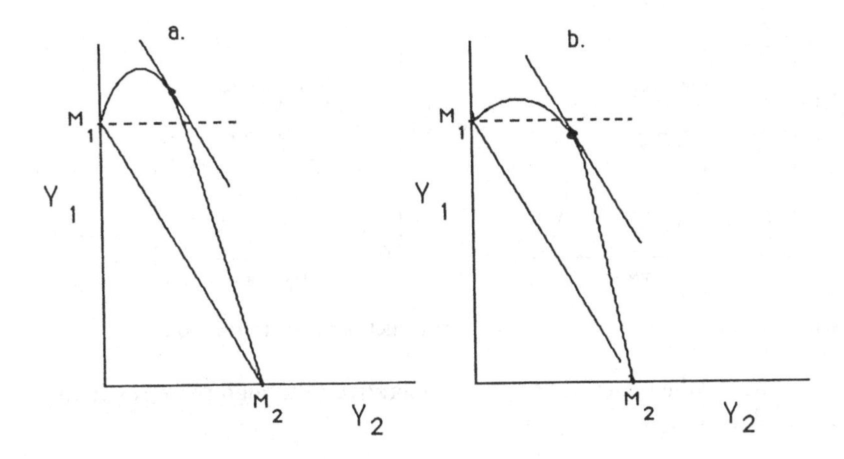

Fig. 4.3. Two examples of yield sets involving facilitation. (*a*) Species 2 facilitates species 1 and the optimal intercrop obtains under the facilitative condition. (*b*) Species 2 facilitates species 1 but the optimal intercrop obtains under the competitive condition (see text).

function of the second species, and that function is g). In Figure 4.2(*a*) the intersection of the two functions is a single one and shows how such an arrangement must result in a point above the unencumbered yield (dotted line) of the species receiving the facilitation. Figure 4.2(*b*) shows how the combined effect of facilitation and competition can produce a nonmonotonic g function. To the left of y^*, species 2 is such that its facilitative effect on the environment is greater than its competitive effect. To the right of y^* the reverse is true. With such a nonmonotonic arrangement it is likely that more than a single intersection will exist. In Figure 4.2(*b*) the alternative intersections imply quite different systems. Point P_1 suggests that species 1 has a relative yield greater than 1.0, due to a facilitative effect felt from species 2, which, in turn, yields at a relatively low level. On the other hand point P_2 suggests that both species are competitively suppressed, yet the system overyields (the point is above the line connecting the two monocultural expectations, indicating $LER > 1.0$).

With regard to the yield set, the facilitative production principle involves only a single change from what has been developed thus far. Some points in the yield set will be greater than the monocultural yields. Figure 4.3 illustrates several examples of yield sets involving facilitation. The set in Figure 4.3(*a*) shows facilitation of species 1 by species 2 and also the optimal system based on the $LER > 1.0$ criterion. In Figure 4.3(*b*) there is likewise facilitation of species 1 by species 2, but here, because of the shape of the yield set, the optimal

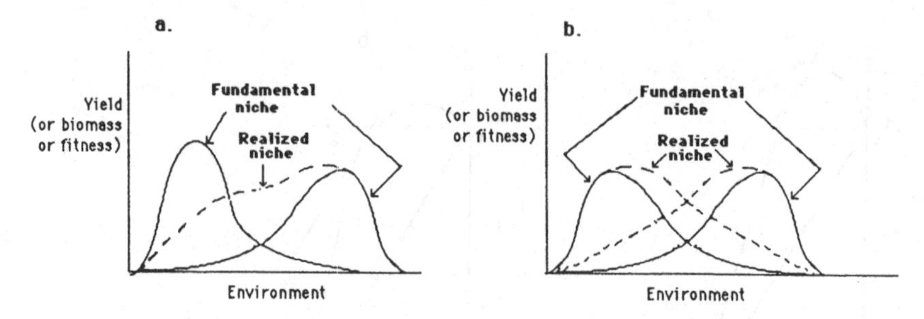

Fig. 4.4. Examples of fundamental and realized niches under facilitation.

system is one in which facilitation is not operative (although there is still an intercrop advantage).

Niche theory

The application of classical concepts of niche theory to questions of facilitation is, as far as I know, nonexistent. The recent summary volume on mutualism (Boucher, 1985) has not even a single reference to niche in its index. But simple common-sense suggests a variety of patterns that are likely to be associated with facilitation. In Figure 4.4 I illustrate two likely patterns. In Figure 4.4(*a*), species 1 is facilitated by species 2 and in Figure 4.4(*b*), both species are facilitators, the condition of mutualism.

It is worth noting that if facilitation is operative, and dominates competition (see below), the general expectation is that the realized niche will be broader than the fundamental niche.

Competition and facilitation together

One of the problems with conceptualizing intercrop advantage as due to reduced competition *or* facilitation is that an automatic asymmetry is implicit. When two plants grow near one another, basic physiological principles suggest that they will almost always compete, whether or not facilitation is operative. To see this relationship, imagine a single plant grown under the influence of a number of individuals of a second species. If competition only is operative, as we increase the density of the second species, the yield of the principal species will decline. A graph of the yield of the principal species versus density of the secondary species will thus be monotonically decreasing.

Now we imagine that competition is totally absent and that some facilitative mechanism is operative. The result will be a monotonically increasing graph of yield of the principal species versus density of the secondary one.

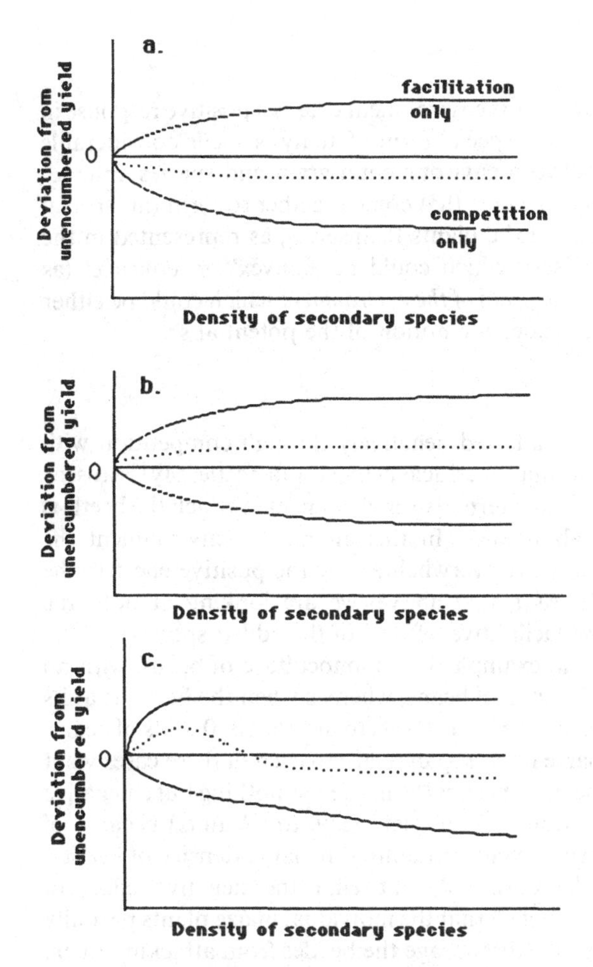

Fig. 4.5. The balance between facilitation and competition (*a*) net effect competitive; (*b*) net effect facilitative; (*c*) net effect competitive or facilitative, depending on density of secondary species.

To see the operation of both together we can recast the above relationships as a graph of change in yield as a function of density. The change in yield will always be negative in the case of competition alone and always positive in the case of facilitation alone. But the net effect could be always negative (Figure 4.5(*a*)), always positive (Figure 4.5(*b*)), or could switch from positive to negative (or vice-versa) as a function of density (Figure 4.5(*c*)). For the practical matter of intercrops we are interested in those cases in which the net effect is positive for at least some densities.

The question of grain

When the effect of one species on the environment causes a positive response in the other species (facilitation), a special form of analysis sheds considerable light on the relationship between environmental grain and species' interactions. There are two distinct concepts that come together to form this special analysis. First, is the biology of the plants themselves, as represented in the *potential set* (not the yield set), which could be 'convex' or 'concave' (as explained below). Second is the *grain of the environment* which could be either 'coarse' or 'fine'. We first consider the notion of the potential set.

The potential set

Suppose that bean yields are affected negatively through competition with maize, in an environment in which a critical bean pest (a beetle, say) is absent. But suppose also that there is an alternative environment in which the beetle is at its maximum possible abundance. In this alternative environment the competitive effect on the beans is overwhelmed by the positive effect of the maize deterring the beetle. As stated above, we are looking at both the competitive 'effects' and the facilitative 'effects' of the added species.

Consider a simple artificial example. In a monoculture of beans, with no beetles present, we expect 20 units of beans, whereas when the beetle is at its maximum density we expect all the bean plants to die, that is, 0 units of beans. If we were to add a very sparse intercrop of maize to each of these cases what would we expect? In the case of no beetles the maize has nothing but a negative effect, so we expect a reduction in bean yield (say, to 19 units) because of competition. But in the environment containing the large density of beetles, the protective effect of the corn will outweigh the negative effect of competition. That is, despite the fact that the individual maize plants partially shade the bean plants, they also discourage the beetles from attacking them. We thus expect an increase in bean yield (say, to 5 units).

Following this general line of reasoning, we might generate a table such as this:

Maize density	0 (monoculture)	1	2	3	4
Yield in Env. 1	20	19	17	14	11
Yield in Env. 2	0	5	8	10	11

To study the yields more closely we can look at the pairs of yields (20,0), (19,5), (17,8), (14,10), and (11,11). That is, for each value of maize density we know of

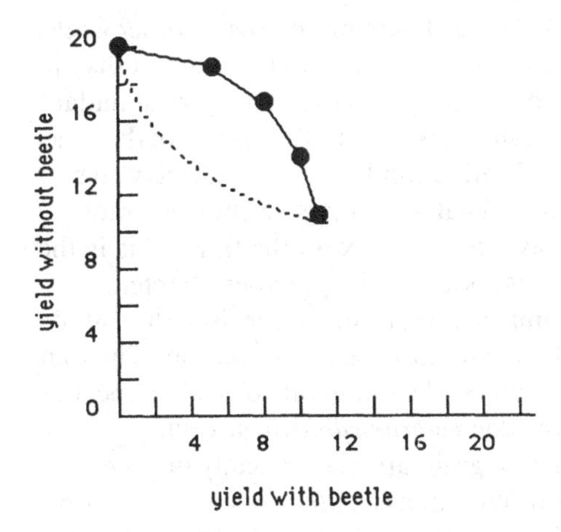

Fig. 4.6. Potential set for example given in text (from Vandermeer 1984b).

two values of bean yield. It is not always necessary to remember the maize density. For some purposes (as here) it makes more sense to look at the coupled yields, that is, each pair of yields. This set of yields is called the 'potential set' (it is *not* the same as the yield set). Actually, the set of all possible pairs, obtained from doing experiments with all possible densities of maize, is the potential set. For our purposes here we can think of the five pairs of yields as the potential set.

A convenient way to study the potential set is by graphing each pair of yields as a point on a graph of 'yield in environment 1' vs 'yield in environment 2'. In Figure 4.6 such a graph has been constructed. The solid line indicates where all of the unknown values would probably lie if the relevant experiments were to be done, and the dots show where the known points are. This graphical representation of the potential set makes a very crucial distinction possible. Suppose the yields had fallen along the dotted line in the diagram, as opposed to where they in fact are. The obvious difference in the shape of the set indicated by the solid line and the set indicated by the dotted line suggests the simple dichotomous classification of convex (the solid line) versus concave (the dotted line) sets. As we shall see later, this distinction is very important in terms of planning an intercropping system.

The environmental grain

In the current example we have two environments (1) very high beetle density, and (2) very low beetle density. From the point of view of the bean plants, how are these two environments viewed? First let us imagine that the beetle density varies from week to week, some weeks being at maximum density, other weeks being at zero density. Since a bean plant lives for about three months, an individual bean plant experiences a sort of fuzzy mixture of the very abundant and the zero density. If the very abundant weeks occur about 50% of the time, from the bean plant's point of view the beetle abundance might just as well be a continuous 50% of its maximal value. That is, a massive (but not mortal) attack 50% of the time is the same as a 50% attack all the time. This is the classic *fine-grained* environment, as discussed in the previous chapter.

At the other extreme, we can imagine that the beetle is either at its maximum or zero abundance, but for an entire growing season. Thus an individual bean plant experiences a high beetle attack rate or a zero one, but nothing inbetween. This is the classic *coarse-grained* environment.

While the two types of environmental grain are diametrically opposed, we can analyze them in a similar fashion. We can ask, 'what fraction of the time will the plant be exposed to each alternative?' In the coarse-grained case the question would be more accurately posed 'what fraction of years, over a period of many years, will the crop experience each alternative?', while in the fine-grained case it would be, 'what fraction of days, over a single growing season, will the crop experience each alternative?' In either case we can talk of the frequency of occurrence of the two environments. Using the example developed in the last section (see above table), suppose that the two environments occur 50% of the time each. The yield over the season, in monoculture will be 10 kg (50% of 20 and 50% of zero). The yield of beans at maize density 1 will be 12 ($0.5 \times 19 + 0.5 \times 5$), at maize density 2, 12.5, etc. The expected yields for each maize density are shown in the following table:

Maize density	0	1	2	3	4
Expected bean yield	10	12	12.5	12	11

So, if we wish to maximize the yield in this environment we would choose maize density 2.

So as to be able to explain other possible combinations, we must generalize the above analysis. To compute the yields with the different corn densities we employed the equation,

$$Y = 0.5y_1 + 5y_2,$$

where Y is the mean yield and y_1 and y_2 are the yields in the two environments. This equation is called the 'adaptive function'. We can rearrange it so that y_1 is expressed as a function of y_2 as follows:

$$y_1 = (1/0.5)\bar{Y} - y_2.$$

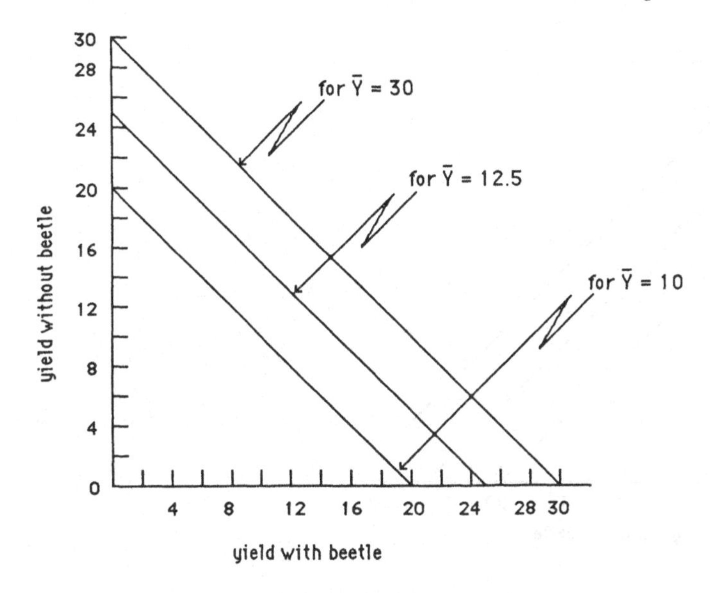

Fig. 4.7. Adaptive functions for various values of y.

In Figure 4.7 this equation is plotted for various values of $\bar{Y}$. Remember that the intent is to maximize the value of $\bar{Y}$. Examining Figure 4.7 we see a value of $\bar{Y}$ that is equal to 30, which would be larger than what we already computed as maximal (i.e., 12.5). The catch is that there are no values in the potential set that can possibly result in the yield of 30. What we're really looking for is the largest value of the yield that is possible, given the constraints of the potential set. We can get a glimpse of how this process operates by plotting Figures 4.6 and 4.7 together. This is done in Figure 4.8. Here we see that the point at which the adaptive function is tangent to the potential set is the point of optimal design for the intercrop. In anticipation of things to come, it is interesting to contemplate what would have been the optimum had the potential set been the dotted line in Figure 4.6.

But what might be the result if, instead of the two environments occurring 50% of the time each, one of them occurred 80% of the time and the other 20% of the time? This time the yield of beans at maize density 1 will be 16.2 ($0.8 \times 19 + 0.2 \times 5$) at maize density 2, 15.2, etc. The expected yields for each maize density are shown in the following table:

Maize density	0	1	2	3	4
Expected bean yield	16.0	16.2	15.2	13.2	11.0

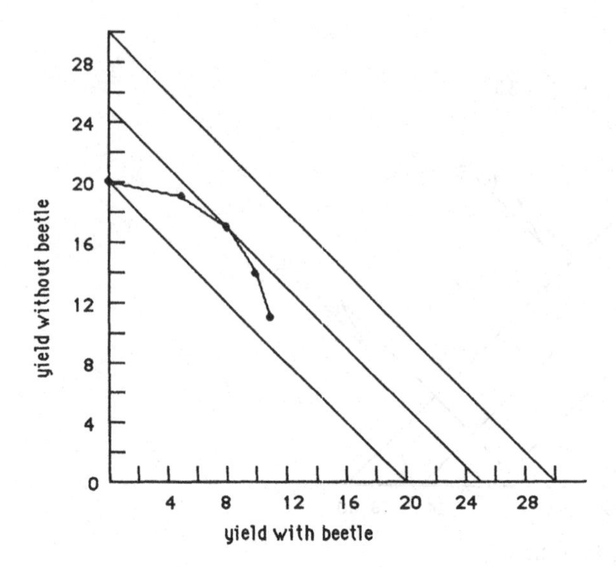

Fig. 4.8. Combination of figures 4.6 and 4.7, illustrating optimization with the adaptive function on the potential set (from Vandermeer 1984b).

Here we see that the maize density would be density 1 rather than as it was before, thus illustrating the important point that as the mix of the environment changes, so will change the optimal design of the system. We now examine this idea in a more rigorous fashion.

More on the potential set

With the background of the previous numerical example, we now move to a general qualitative discussion of what the potential set is all about. This is intended as a generalization of the specific example given above, but not a true formalization, which will be presented subsequently. We consider the situation in which one crop, the 'secondary crop', alters the environment of a 'principal crop'. As the density of the secondary crop increases, we would usually expect an increasingly negative effect felt by the principal crop due to competition from the secondary crop (Harper, 1977; Silvertown, 1982). But we also assume that some other effect, a facilitative effect, is present (e.g. enrichment of nitrogen environment, protection from insects). It is not that the facilitative effect must occur in all intercrops, it certainly does not. Rather, we are here concerned with those cases when it in fact does.

Next we conceive of two extreme environments, one in which only the

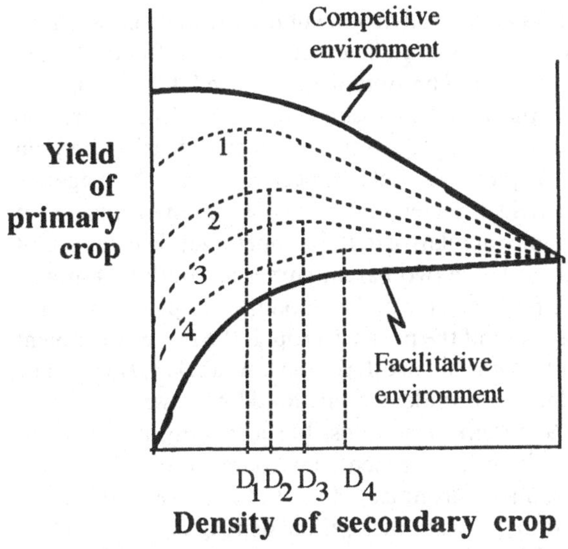

Fig. 4.9. Yield of primary crop as a function of density of secondary crop in alternative environments. Dotted lines illustrate different mixtures of environments, as explained in text (from Vandermeer 1984b).

negative (competitive) effect of the secondary crop is realized and an alternative one in which the positive (facilitative) effect is strongly realized. For example, if the positive effect is total protection from host-specific insect pests, one environment of the principal crop would be total absence of the insect, in which case only the negative effects of competition from the secondary crop will be realized. The other environment would contain a very large population density of the pest, in which case the positive effect of competition from the secondary crop would be predominant. Or, to take another example, suppose the positive effect is an augmented supply of nitrogen, obtained from biological nitrogen fixation in a legume secondary crop. One environment would be saturated with nitrogen, thus negating the potential facilitative effect of the legume, and rendering the net effect of the secondary species negative. The alternative environment would have very low nitrogen levels, in which the positive effect of the legume secondary crop would be most pronounced.

In general, a graphical plot of the yield of the principal crop versus the density of the secondary crop will be increasing for the environment that favors the positive effect (the 'facilitative' environment) and decreasing for the environment that does not (the 'competitive' environment). Such a plot is presented in Figure 4.9.

Next we consider what might be the relationship between principal crop yield and secondary crop density in an environment intermediate between the competitive and facilitative environments. The dotted lines in Figure 4.9 are meant to represent that situation. The dotted line labeled 1 in Figure 4.9 represents a situation where the actual environment is 75% of the theoretical facilitative environment. That is, at any given secondary crop density, the yield of the primary crop is 75% of the yield expected under the negative environment relative to the yield expected under the positive environment (or 25% of the facilitative environment relative to the competitive). The important feature of this curve is that while the two parent curves are both monotonic, the intermediate curve has a peak, a value of the secondary crop density which results in a maximum of the yield of the primary crop. If the real environment were as stipulated by the dotted line labeled 1, the density labeled D_1 would be the 'optimal' density at which a secondary crop should be sown.

Also shown in Figure 4.9 are three other possible environments. Environment 2 is 50% of the competitive environment, environment 3 is 25% of the competitive environment, and environment 4 is 10% of the competitive environment. It is of interest to note that the optimal density of the secondary crop changes as a function of the actual environmental pattern, from D_1 to D_2 to D_3 to D_4.

The rest of this chapter is aimed at the development of certain mathematical formalities which enables a more detailed examination of the kind of phenomenon set out in a simplified qualitative form in Figure 4.9. The model that follows draws heavily on the work of Levins (1962, 1968), but with a substantial change in the biological interpretation of nonlinearities in the decision process, most of which appears explicitly in Chapter 9.

The potential set, adaptive function, and environmental grain – the details

What follows is not really difficult, but it is somewhat complicated. The reader interested in only a general qualitative idea of the nature of the facilitative production principle might consider skipping the rest of the chapter, although many interesting features of the principle cannot really be explained in a more simplified manner and thus might not be fully appreciated by the reader who skips. Following my earlier work (Vandermeer, 1984*b*), the development is divided into four sections: (1) the range of systems and the potential set, (2) the adaptive function, (3) maximizing yield on the potential set, and (4) the shape of the potential set.

The range of systems and the potential set

We conceive of a series of possible systems ranging from a monoculture of the principal crop to a system of the secondary crop that is very intense, what is called in this chapter the 'full intercrop'. Usually this gradient will correspond to ever more concentrated plantings of the secondary crop. For example, with a maize–bean system in which the maize is the principle crop and the bean the secondary one, the system would range from a monoculture of maize, through a system with a few scattered beans, through a system where beans are as abundant as maize, to the case where beans are so dense that they are totally dominant.

Certain conceptual difficulties are immediately encountered in trying to order the potential systems in this fashion. What densities of maize are used in the various systems? Are the systems replacement series? These and other questions are undoubtedly important and must be dealt with in any practical application. But for the development of the theoretical generalization one only has to assume that it is possible to order potential systems along a gradient which makes some kind of sense. The specification of this ordering process may be different for different intercrops, but the assumption is made that for any given combination, some form of ordering will be possible in which there is an inherent quantitative 'sense' to the ordering (e.g. increasing bean density) and that it results in an increase in levels of facilitation for the principal crop species, at some points on the gradient, in the facilitative environment.

Next recall the two extreme environmental circumstances, one in which only the competitive effect is felt, the other in which the maximum possible facilitative effect is felt, described earlier as a competitive environment and a facilitative environment. The overall conceptual scheme can be conveniently represented on a single graph, as was done in Figure 4.9.

The functional representation of the yields in the two environments can be presented in its dual form, by plotting each system as a point on a graph of Y_{I} versus Y_{II}. This is what was done in the artificial example above, and is illustrated in Figure 4.10. Associated with each potential density of the secondary crop (i.e. with each point on the abscissa in Figure 4.10) are two potential yields, one associated with the competitive (Y_{I}) and one associated with the facilitative environment (Y_{II}). These two yield values may be plotted on a graph of Y_{I} vs Y_{II} (see Figure 4.10) for all points on the abscissa of the original graph. The resulting set of points, representing all potential combinations of Y_{I} and Y_{II}, is referred to as the potential set (*not* the same as the yield set discussed earlier).

The adaptive function

The mere existence of the potential set is not interesting until we add the notion of the adaptive function. Suppose the factor that is being modified in the environment is continually there, as opposed to only occasional occurrences, and that it operates at some points in a field at full intensity and other points at

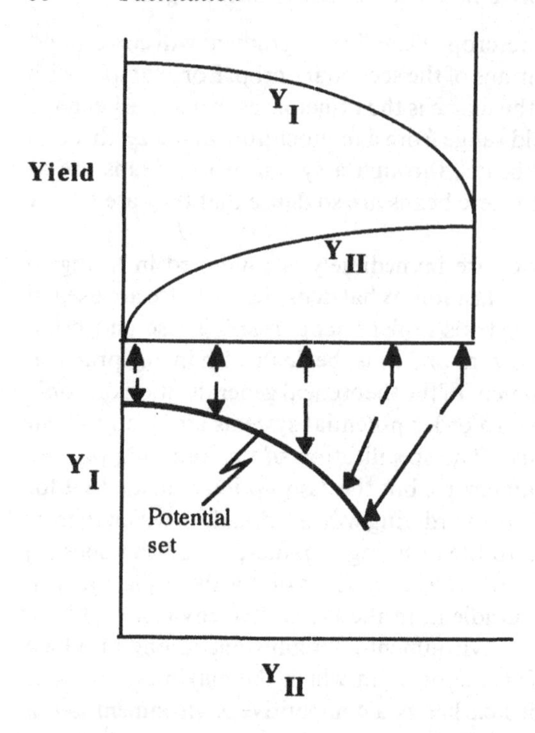

Fig. 4.10. Illustration of the construction of the potential set from the yield versus density relation. Top graph represents the yield of the primary crop in environment I (Y_I) and in environment II (Y_{II}) as a function of density of secondary species (the abcissa). Each point on the abcissa represents two yields, one in environment I (Y_I), the other in environment II (Y_{II}). These pairs of points can be plotted on a graph of Y_I versus Y_{II} to form a potential set (from Vandermeer 1984b).

minimal intensity. So, for example, we might have patches of high nitrogen and low nitrogen, or patches of high and low pest attack, or patches of high moisture or low moisture. One patch represents the competitive environment and the other patch represents the facilitative environment. Then, if p is the fraction of the field in which the pest is absent, the expected yield of the principal crop in the field would be

$$Y = pY_I + (1-p)Y_{II} \tag{4.1}$$

where Y_I is the yield of the principal in the competitive environment and Y_{II} its yield in the facilitative environment. Equation (4.1) is one form of what we here define as 'the adaptive function'.

It is convenient to think of the environmental situation as a fraction, p, of the field under the influence of the competitive environment and another fraction, $(1-p)$, under the influence of the facilitative environment. But, the true meaning of equation (4.1) is not nearly so restrictive. The actual yield in the field (Y), will be greater than the minimum (Y_{II}) at any density of the secondary crop. With this density, the difference ($Y-Y_{II}$) between the actual and minimum must be some fraction of the difference (Y_I-Y_{II}) between the maximum and the minimum. If that fraction is p, we have

$$Y-Y_{II}=p(Y_I-Y_{II}),$$

which, after some algebraic manipulation, can be seen to be equivalent to equation (4.1). The meaning of the fraction p is, the ratio of the actual yield above minimum, to the potential above minimum, or $p=(Y-Y_{II})/(Y-Y_{II})$. Nevertheless, it bears repeating that it will be much more convenient, for heuristic purposes, to think of p as the fraction of the field under the competitive environment and $(1-p)$ as the fraction of the field under the facilitative environment.

The pattern described above is a fine-grained pattern, as was earlier noted, in which two (or more) environmental patterns occur simultaneously from the point of view of the organism experiencing them. This pattern is to be contrasted to a coarse-grained pattern in which first one, then the other (and then others, if there are more than two) environment is exposed to the organism. With a coarse-grained environment the environmental alternatives are viewed one at a time, rather than in a mixture (as was the case in the fine-grained environment).

With regard to the problem at hand, some environmental alternatives clearly present themselves in a coarse-grained fashion; for example, 'plague'-type insect pests, devastating crops in one year and being absent in the next, such that a facilitative secondary crop would be one that offered the primary crop some protection against the pest. Here we conceive of the fraction of years that the environment is composed of the competitive alternative as p and the fraction of years that the environment is composed of the facilitative alternative as $(1-p)$. Or if they are spatially distributed alternatives, such as patches of wet areas in a newly cleared forest, the fraction of patches that are competitive is p and the fraction that are facilitative is $1-p$.

The average yield over the whole region or over the long term (if it is a time distributed environment) will simply be

$$Y=pY_I+(1-p)Y_{II}$$

which is exactly the same as equation (4.1). If the producer wishes to maximize yield, the adaptive function which should be considered is the same for both a coarse and fine grained environment. In Chapter 9 it will be seen that when the decision criterion is altered, different environmental grain leads to qualitatively different adaptive functions.

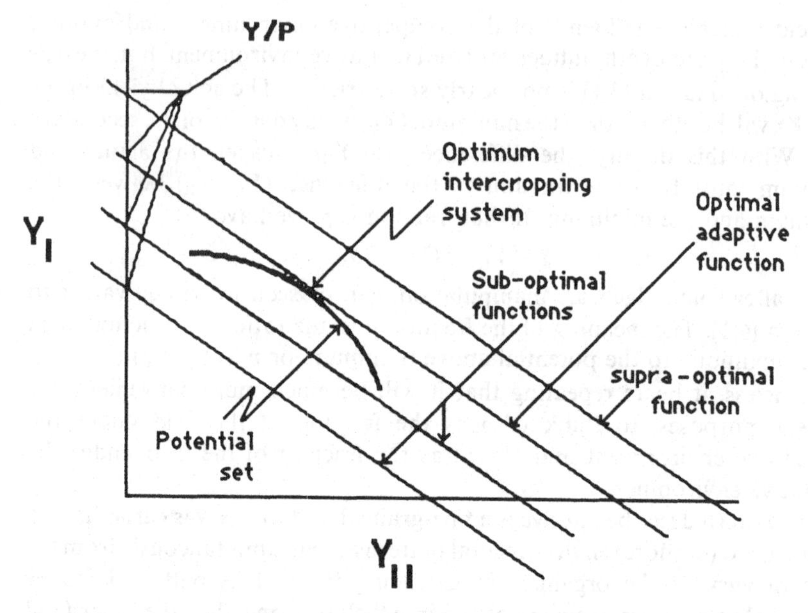

Fig. 4.11. Optimizing the intercrop with the adaptive function on the potential set (from Vandermeer 1984b).

Maximizing yield on the potential set

Proceeding now to the utility of the adaptive function, we may algebraically alter equation (4.1) so that it reads

$$Y_{\mathrm{I}} = (Y/p) - [(1-p)/p]\,Y_{\mathrm{II}}.$$

From this we see that the adaptive function can be viewed on a graph of Y_{I} vs Y_{II}, in which case it will appear linear with slope $(1-p)/p$, and intercept Y/p. Since Y is to be maximized, the maximization procedure is easily visualized as increasing the intercept of the adaptive function (i.e. increasing Y/p, equivalent to increasing Y alone since p is constant) while keeping its slope constant.

But exactly how far we are able to push Y/p is dictated by the potential set. As its name implies, the potential set represents all possible systems. Thus we seek the adaptive function that has the largest value of Y, but that also intercepts the potential set at one (or more) points. Viewing the system in this manner allows us to find the optimum intercrop, at least qualitatively, from a superficial examination of the potential set and the adaptive function, as illustrated in Figure 4.11 which illustrates a potential set and a series of adaptive functions, one of which is clearly optimal. The others are either suboptimal, in that they represent a value of Y which is lower than the Y obtained with the optimal adaptive function, or supra-optimal in that it

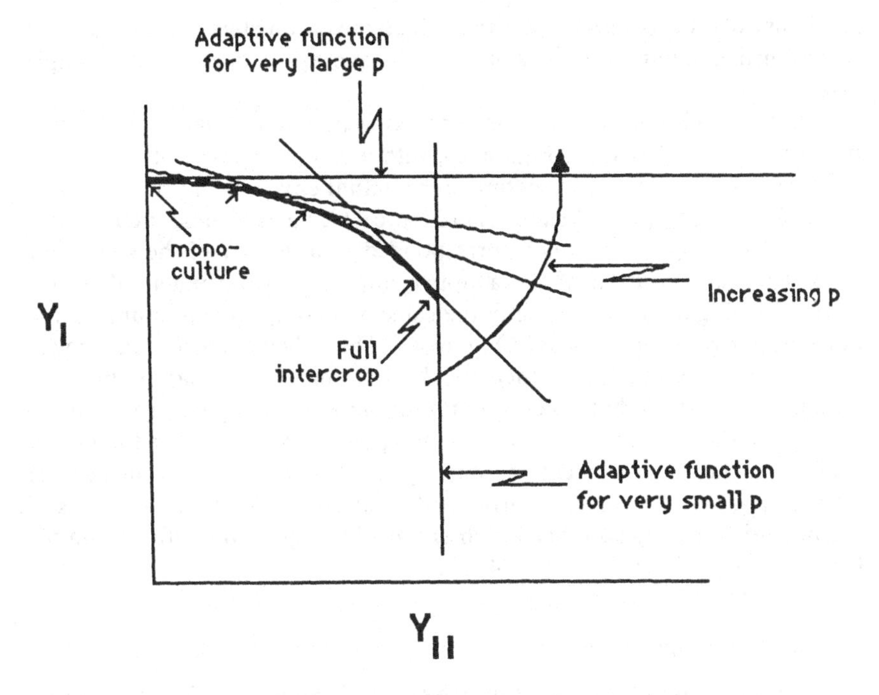

Fig. 4.12. Changes in the optimal solution as the environment changes. Each small arrow pointing to the potential set indicates the position of an optimal point corresponding to one of the adaptive functions (from Vandermeer 1984b).

represents a value of Y which is higher than that represented by the optimal, but does not intersect the potential set, which is to say there is no possible system which will give the yield.

In Figure 4.11 an intermediate intercrop will be the optimum, that is, the adaptive function is tangent to the potential set at a point which represents a system intermediate between the monoculture and the full intercrop. But we might also ask what would happen if some external force altered the environment, say either *Homo sapiens* or natural accident. Such an alteration can be viewed as a change in the value of p.

Since the slope of the adaptive function is $(1-p)/p$, a change in p will change the slope. If p is decreased, the adaptive function approaches a vertical line, and the optimal system is the most intense application of the secondary crop. As p increases, the gradient decreases, and the optimum adaptive function moves along the potential set, which is to say the optimal system becomes more like the monoculture. Eventually, the monoculture is the optimal system. This process is illustrated in Figure 4.12. Qualitatively, we see the not unexpected result that as the facilitative environment becomes less frequent, the use of the secondary plant becomes less desirable. If all the parameters of the model can be specified, computation of the optimal mix of crops is, in principle, a simple task.

This process is precisely the same process, viewed in its dual form, that was presented earlier. Recalling Figure 4.9, with a certain fraction of each of the extreme environments representing some actual environment, intermediate curves were drawn (the dotted lines in Figure 4.9). Each of those intermediate curves had an optimum which corresponded to a density of the secondary crop. The maximization of the adaptive function on the potential set (as pictured in Figure 4.11) is exactly the same as finding the optimum for the intermediate environments as in Figure 4.9. The only difference is in the way the problem is approached mathematically. Maximizing the adaptive function on a potential set perhaps seems a bit esoteric when compared to the more obvious method of Figure 4.9. But it is an approach which allows for a good deal of further development, permitting us to examine consequences of different and more complex environmental patterns and alternative forms of the potential set, a possibility which is out of the question with the simple formulation presented in Figure 4.9.

The shape of the potential set

Thus far, all examples presented in this section have been with potential sets that bow outward, that is, with convex sets. If the shape of the set is different, certain results change rather dramatically.

The construction of the potential set, as pictured in Figure 4.10, has so far been presented for two yield curves which themselves both bow outward, and are thus convex. It is certainly possible that both yield curves could bow inward, that is, be concave. If this were the case, the potential set itself would be concave, as is the one pictured in Figure 4.13.

Following similar arguments as in the last section, if we begin with a very large value of p (probability of facilitative environment very small), the adaptive function is almost a horizontal line and the optimum is a monoculture (see Figure 4.13). Following the same reasoning as before – which led to a gradual shift in the optimal intercropping system – we allow p to become smaller. But in the case of a concave potential set we do not observe a gradual change in the optimal system. Rather the optimal system remains a monoculture over a broad range of values of p (see Figure 4.13(*a*)). However, as p continues to decrease further, a critical point is reached in which both the monoculture and full intercrop are equally optimal (see Figure 4.13(*b*)). Further decreases in p beyond this critical value result in the highly intense secondary crop presence (the full intercrop) being optimal (Figure 4.13(*b*)). Thus, contrary to the results of changing p with a convex potential set (a

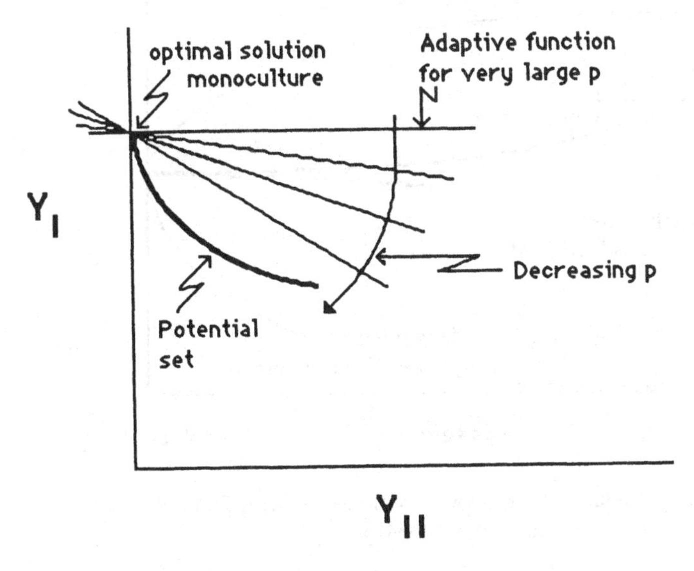

b.

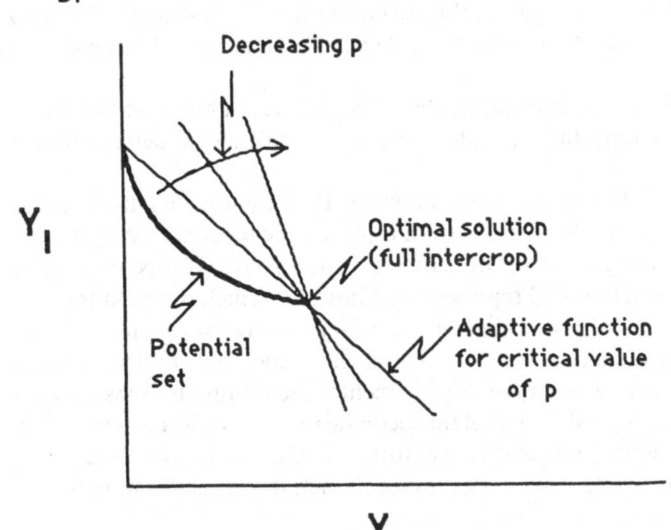

Fig. 4.13. The optimal intercrop under conditions of a changing environment but with a concave potential set. (*a*) As p decreases the optimum remains the monoculture. (*b*) At a critical point the optimum switches to the full intercrop (from Vandermeer 1984b).

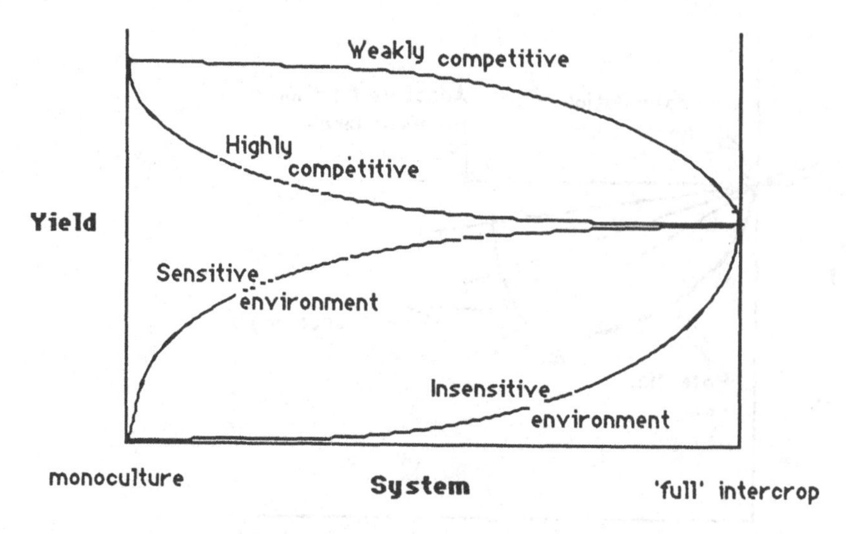

Fig. 4.14. The various relationships between yield and system that result in potential sets (see text) (from Vandermeer 1984b).

gradual change in the optimal intercropping system), effecting the same changes in p with a concave set, results in a constant optimal strategy, followed by a sudden shift to another optimal strategy, as illustrated in Figure 4.13(*a*) and (*b*).

Given such drastic differences in the consequences of concave vs convex potential sets, it is important to understand, biologically, what determines the shape of the set.

Looking first at the yields expected when the facilitative environment is present (Figure 4.9), we have a situation in which the secondary crop has an immediate and strong effect on the environmental factor. If this yield curve were to bow inward, it would represent a situation in which the yield remains relatively small over a broad range of secondary crop densities but increases rapidly at higher densities. In other words the convex yield curve, as first described, represents a sensitive environmental factor (in the sense that it responds to even sparse plantings of the secondary crop), while a concave yield curve would represent an insensitive environmental factor (in the sense that it does not respond to the secondary crop until high densities of the latter are reached).

Turning to the case of the competitive environment, we are concerned with the competitive effect of the secondary crop on the primary crop. Again making reference to Figure 4.9, when the yield curve is convex we are dealing with a weakly competitive secondary crop. If the yield curve is concave, the secondary crop is highly competitive.

These four qualitatively distinct forms are illustrated in Figure 4.14. Now we are able to identify two extreme situations. First, we consider a sensitive environment in combination with a weakly competitive secondary crop, a combination which inevitably leads to a convex potential set. Second, at the other extreme, we consider an insensitive environment in combination with a highly competitive secondary crop, a combination which inevitably leads to a concave potential set.

Thus, the whole framework here can be summarized as a yield response of a principal crop to an intensity (probably usually the density) of a secondary crop, in two alternative environments, a competitive one and a facilitative one. This response can be formulated in terms of a potential set which has either a convex or concave form. The pattern of the environment, which is viewed as particular combinations of the competitive and facilitative alternatives, can be represented as an adaptive function. The combination of the potential set and the adaptive function stipulate how one may find the optimum intensity of the secondary crop.

5

Mechanisms of the competitive production principle

The competitive production principle, as elucidated in Chapter 3, in a sense needs no further explanation or mechanism to explain intercrop advantage. It is just the case that the intercrop system is advantageous over its monocultural alternatives if competition between the two crops is not very intense. But it nevertheless makes sense to ask, at a more mechanistic level, what is the mechanism whereby competition is rendered small. This is a common question in community ecology, where one uses phrases like 'niche subdivision', 'habitat partitioning', 'niche overlap', 'resource partitioning (or overlap or subdivision)', and probably others to refer to the same set of phenomena. When two species are obviously doing more or less the same thing in an environment, as described in Chapter 3, it is natural to expect one or the other to dominate, much as competition in sports results in a winner and loser. As we now know, stalemate in nature is as common as checkmate. When expectations of a winner and loser are assumed because of obvious similarities in the two species, but they nevertheless coexist in the same environment, it is most natural to ask why. Why is the expected extinction of one of the species not observed? Formally the question is usually posed as 'what is the mechanism of coexistence?'

Given the formal similarity between the competitive exclusion principle and the competitive production principle, it makes sense to ask 'what is the mechanism of intercrop advantage?' even if it is known that the advantage stems only from the weakness of the competitive interaction. There is something of a philosophical point to be recognized here, although it is clearly beyond the scope of this book. How do we recognize a situation in which it makes sense to question the mechanism of coexistence or intercrop advantage? One would hardly be concerned with the problem of explaining how lions and acacia trees are able to coexist, or why small farmers raise chickens in the same yard with their fruit trees, yet such problems are formally identical to searching for the mechanism of coexistence or intercrop advantage. We tacitly assume that the informal information we have about the two species is sufficient to cause us to expect one to force the other to extinction or one to gain the yield advantage over the other. When, in the face of these expectations, the two species nevertheless coexist or demonstrate a yield advantage, it becomes necessary, only because of that tacit assumption, to devise an explanation as to why. What is the mechanism? What is the mechanism which results in the opposite of our expectation, coexistence where we expected extinction, and intercrop advantage where we expected none?

While a moment's reflection might suggest that the more interesting question is why did we come to the original assumption in the first place, such philosophical questions are better left to other forums. For now we accept the popular and conventional wisdom that when, for however capricious the evidence, we accept the proposition that two crops grown together should not show an advantage over two crops grown apart and they nevertheless overyield, the mechanism as to how they do this is an interesting subject for study. What is the mechanism of overyielding, assuming that the competitive production principle is operating?

Ecologists have traditionally approached this question at a high level of abstraction. The vehicle of the ecological niche has, in a sense, bridged the gap between the questions 'do they coexist?' and 'what is the mechanism of their coexistence?' The idea is, as already expressed in Chapter 2, if two species coexist (or, equivalently, two crops demonstrate an intercropping yield advantage), it is likely that they do so because their niches do not overlap sufficiently. The tacit assumption here is that the degree to which niches overlap is a measure of the intensity of competition between the two species (Colwell & Fuentes, 1975), an assumption that is quite obviously dependent on how one measures the niche overlap. It is essential to remember that the transformation problem from niche overlap to competitive intensity is an assumption, probably sometimes valid, certainly at other times not valid. It may be, nevertheless, that one may gain some insight into an intercropping system by conceiving it as two species living along one or several resource gradients, bearing in mind the severe problems.

This conceptual framework, widely used in ecology, has not proved to be very useful. It is little more than a restatement of the competitive exclusion (production) principle in the first place, and requires the transformation assumption (niche overlaps are akin to competitive intensity) to be true. It has thus probably led to more confusion than enlightenment, especially when it has been employed to try to quantify the concept of competition coefficient (Colwell & Fuentes, 1975; Colwell & Futuyma, 1971; Culver, 1970, 1985; Schoener, 1968). As long as it is used as a qualitative tool for reformulating the competitive production principle I suppose no harm can be done. Nevertheless, a great deal of caution and skepticism should be exercised in any attempt to quantify the overlap and relate that figure to the intensity of competition, as has already been discussed in Chapter 3.

However, formulating the problem in this manner does force us to consider what might be the resource gradients involved in the competition process, which after all is the whole meaning of the competitive production principle. This classic question, in the case of plants, can be divided conveniently into two themes, competition for light and competition for the things that occur in the soil. While this is not the only possible categorization (e.g. see, Weiner & Thomas, 1986; Horn, 1971; Harper, 1977), it is the most obvious (to me) and probably the most popular. It is thus used in this chapter.

Partitioning the light environment

Probably the major problem in researching questions of the light environment, derives from truly severe technical problems. As Harper (1977) remarks: 'The technical limitations in the measurement of light within vegetation have been reviewed by Anderson (1964, 1966) and after reading her papers it is the brave man [*sic*] who researches in this area'. We nevertheless procede with some theoretical comments, cautious of but not cowed by, the technical difficulties of measurement.

It is common knowledge that the rate of photosynthesis is an increasing function, with diminishing returns, of the intensity of light, that is, the rate of photosynthesis increases rapidly when a low level of light is elevated and increases slowly when a high level of light is elevated. But it is not, as might initially be assumed, that a faster photosynthetic rate results in competitive dominance. It may be true for very similar organisms (e.g. two varieties of corn or two phytoplankton), but certainly not generally. Examples are too numerous and obvious to mention (climax tree species replacing pioneer species, anything with allelopathy, tall, broad-leaved weeds versus corn). It is for this reason that ecologists long ago insisted on the separation of the competitive process as a whole into two (at least) components, rapidity of growth (which undoubtedly *could* lead to competitive dominance) and 'competitive ability', or the effect that a unit biomass (or other measure) of one species has on the other.

In many ecology textbooks these two concepts are presented in the form of the Lotka–Volterra equations, a tradition that is certainly not necessary, but continues to serve heuristic purposes. Here we are concerned with the construction of competition coefficients, conceptually, from principles of light interception. To do so requires two simple observations. First, photosynthesis shows a regular pattern of increase with 'diminishing returns' as a function of light intensity. Second, the peculiarities of different species' physical architecture imply peculiarities in their actual light-intercepting properties.

Figure 5.1 shows photosynthetic curves for two theoretical crop species. The first is a 'sun species' and the second a 'shade species', categorizations that

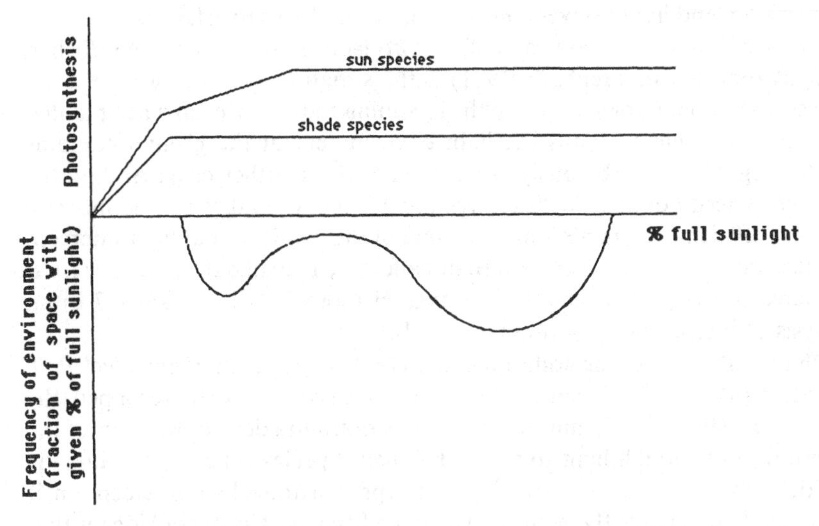

Fig. 5.1. Illustration of an ideal light environment (lower graph) for incorporating a sun and shade species as an intercrop.

have been suggested many times in the literature (e.g. Black *et al.*, 1969), if not exactly by these specific terms. It would seem obvious that when these two species are intercropped, if the light environment created by the combination has peaks at the points where each of the two species are photosynthesizing at sufficiently high rates, the intercrop will obviously be advantageous, since each of the species will be photosynthesizing at the same rate they would have been in a monoculture. But how exactly is the light environment created by the two?

Two species growing together form a canopy that intercepts light qualitatively and quantitatively differently than either of the monocultures. In particular, Trenbath (1981) has developed this idea quite fully. Trenbath analyzes the light use efficiency (LUE) of an intercropping system. The light use efficiency is the light-conversion efficiency (the quantity of intercepted light actually used for photosynthesis) multiplied by the proportion of light intercepted.

Expressing these concepts mathematically we note that the proportion of light intercepted is I_i/I_0, (where I_i is intercepted light and I_0 is light above the canopy) while the light-conversion efficiency is P_n/I_i (where P_n is net photosynthesis), which gives us

$$LUE = (I_r/I_0)\,(P_n/I_i) = P_n/I_0. \tag{5.1}$$

Expressing the light use efficiency (the net photosynthesis as a proportion of incident radiation) as the product of these two factors (proportional light interception, and light-conversion efficiency) enables us to ask questions as to the effect of intercroping on each of the two factors. At one extreme (Fisher, 1975, as reported in Trenbath, 1981) is the situation in which a taller crop species does not completely utilize the incoming radiation, even when planted at its optimal density. Thus the light environment at the ground contains 'wasted' light, which obviously could be used by another crop. Adding the other crop need not affect light conversion efficiency at all, yet could increase light use efficiency by increasing the value of the proportional light interception (i.e. the second crop species need not affect P_n/I_i at all but could alter *LUE* by increasing $I_i/_o$). This situation is treated more fully in Chapter 7 in an analysis of intercrops involving perennials.

On the other hand, the addition of a second crop species could affect P_n/I_i directly. For example if a sun species and shade species are intercropped, the sun species, almost by definition, when at its optimum density will permit the penetration of enough light to support a shade species beneath it. While the addition of the shade species will change the proportional light interception, it is also likely to change the light-conversion efficiency. The overall idea is that an intercrop could be designed in the same way that Warren Wilson (1961) originally suggested a pasture should be, namely with leaf inclinations at the top of the canopy inclined so that they intercept only enough light to photosynthesize maximally, increasing their inclinaion as they are positioned lower in the canopy. Thus each leaf is angled so that it intercepts only the amount of light that it needs. Clearly, if the first species is a sun species, the level in the canopy in which the leaf is no longer above its compensation point is the level at which enough light will still be coming through the canopy to be able to position a shade species.

Applying this concept can be a somewhat tricky process and has yet to be fully worked out, since the growth of the plant through time provides a variety of light regimes. Nevertheless, as noted by Trenbath, the idea is clear. Adding a second crop species might easily be accomplished in such a way that the light-conversion efficiency is maximized, but at a very low value of proportional light interception. Theoretically, one could then alter the leaf angles to the point where the disadvantage of a low light-conversion efficiency is balanced by a relatively high proportion of light interception.

As elegant as Trenbath's approach may be, a slightly simpler approach may be appropriate for some circumstances. In particular, the schematic analysis of Donald (1961) has been useful for monocultural situations and, it seems to me, could be expanded to include intercrops. While this approach obviously ignores many important and interesting factors, it is nevertheless potentially useful as a general theoretical focus. In Figure 5.2 Donald's diagrammatic representation of an idealized grass species is redrawn. Each leaf layer contains two leaves photosynthesizing at a rate appropriate to the light conditions of

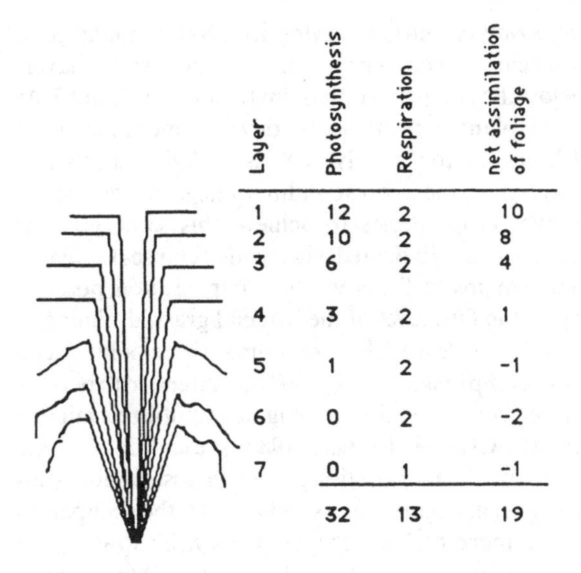

Layer	Photosynthesis	Respiration	net assimilation of foliage
1	12	2	10
2	10	2	8
3	6	2	4
4	3	2	1
5	1	2	-1
6	0	2	-2
7	0	1	-1
	32	13	19

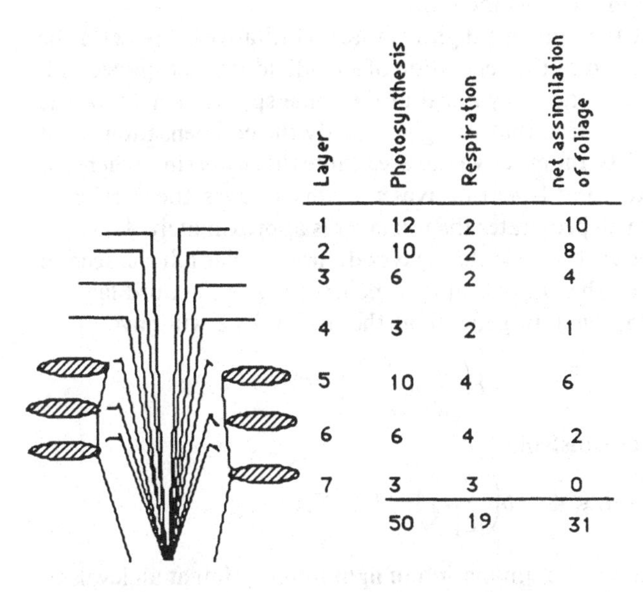

Layer	Photosynthesis	Respiration	net assimilation of foliage
1	12	2	10
2	10	2	8
3	6	2	4
4	3	2	1
5	10	4	6
6	6	4	2
7	3	3	0
	50	19	31

Fig. 5.2. Incorporation of a shade species (shaded elipses) below a sun species, such that the former takes the place of the latter's respiratory drain leaves (modified after Donald, 1961).

that layer, and respiring at a rate of unity per living leaf. Net assimilation is shown in the last column and can be seen to be positive for the first four layers but negative (light levels below the compensation point) at levels 5, 6, and 7. As an evolutionary strategy all plants are likely to develop mechanisms to maximize the number of leaves at the positive net assimilation levels and minimize the number of leaves at the correspondingly negative levels. An intercrop simply uses another crop species to achieve this same goal, as illustrated in Figure 5.2(*b*). Here I have indicated with shaded ellipses the leaves of some shade species that remains well above the compensation point at shade levels corresponding to the fifth level of the original grass, declining in photosynthetic capacity in lower levels, but retaining the shade-species characteristic of relatively low compensation point. What is intended to be clear from Figure 5.2 is that an intercrop can easily be imagined to utilize available solar radiation more efficiently. Clark & Francis (1985) present data on the distribution of vegetation in two bean varieties grown in association with maize. Their data are reproduced in Figure 5.3. As can be seen, the occupation of space above the ground is more or less complementary, illustrating the principles discussed above. It is also apparent from this diagram that a greater advantage should accrue to the combination with bush bean rather than climbing bean, which in fact was the case.

These simple ideas can be put in a slightly more quantitative form. Let A_1 be the ground area subtended by the vegetation of an individual of crop species 1, and let a be the area subtended by a leaf of the same species. Let P_1 be the fraction of incoming radiation that will give exactly the compensation point for this species and let b be the effective leaf area (note that $b = a \sin \theta$ where θ is the leaf angle). If the topmost layer of leaves contains n leaves, the fraction of incoming radiation which penetrates the top layer is approximately $A_1 - n_1 b_1$ (for heuristic purposes we presume that leaves do not overlap). If the second layer contains n_2 leaves, the amount of light penetrating to the third layer is $A - n_1 b - n_2 b = A - b(n_1 + n_2)$. In general, at the ith level we will have

$$A_1 - b\left(\sum_{j=1}^{i} n_j\right).$$

Define the mth layer as satisfying

$$p_1 \leqslant A_1 - b\left(\sum_{j=1}^{m} n_j\right) = A_1 - N_m,$$

where N_m is shorthand for the summation of light interception at all levels to m. If this species is a sun species, we expect p to be relatively large. Consider now a shade species, for which $p_2 < p_1$. Thus if $p_1 < A_1 - N_m$, layers of leaves at level m and possibly below, will produce above their own compensation point. It would thus make sense to place the second species in this position, rather than

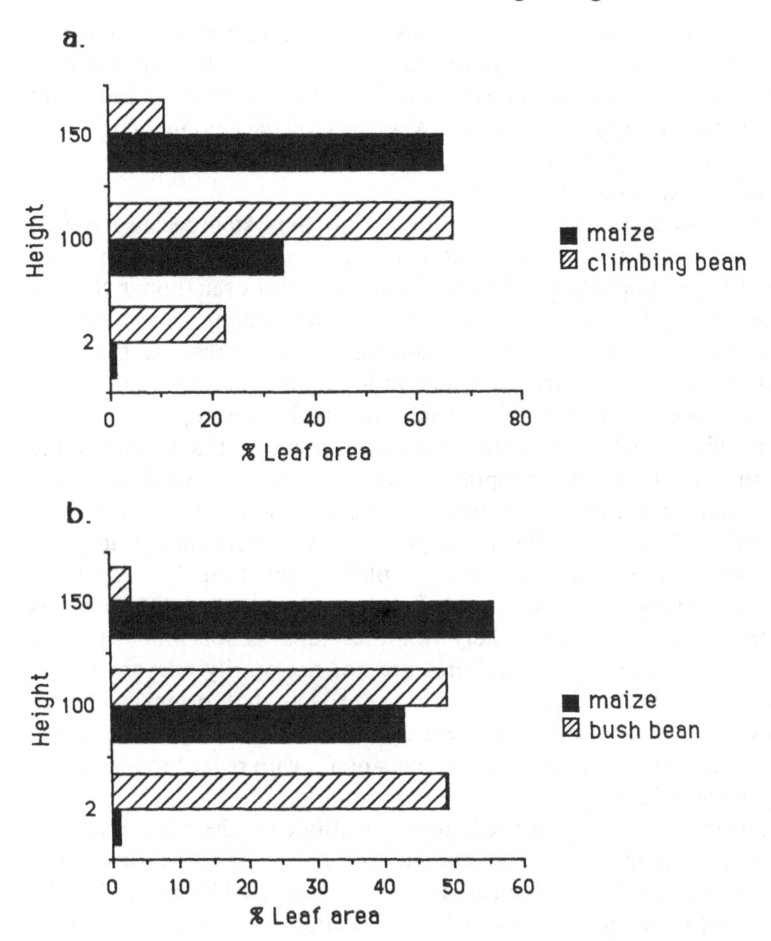

Fig. 5.3. Percentage leaf area occupying different strata in a maize bean intercrop, (data from Clark & Francis, 1985).

more leaves of the first species which would be nothing more than a respiratory drain on the system as a whole.

Assuming that leaves do not overlap makes the above argument quite easy to follow, but it is not a valid assumption. Most workers agree that leaf overlap leads to some sort of nonlinear relationship between light interception and either height in the canopy or leaf area index. For example, Beer's law (Chang, 1968; Monsi & Saeki, 1953) holds that light interception is a negative exponential function of the leaf area index. On the other hand, Stanhill (1962) empirically found a direct linear relationship between light intensity and crop height. While such details are of considerable interest in engineering intercrops, and are further considered in later chapters, the only point of importance here is that from the top to the bottom of a crop, canopy light available for photosynthesis diminishes. At some level the amount of light falls below the crop's compensation point. At that point a crop with a lower compensation point could be put in.

In the above example, the second species fits into the system because the first falls below its compensation point. But there will frequently be cases in which the second species should be introduced at a specific level, even though the first species has not yet reached its compensation point, recalling criteria such as LER and RVT (Chapter 2). If maximizing photosynthesis is the goal, intercrops will sometimes meet that goal more easily than monocultures.

But in practise the process of determining which intercrops combine to produce an efficient light-interception machine is not at all that simple. In the actual planning of an intercropping system we are not concerned with individual plants but, rather, with populations. Thus when $A - N_m$ (or some more complicated function) refers to the proportion of light penetrating to the mth layer, we are referring to a particular plant population. As population density increases so will N_m, and, considering that optimal population densities for intercrops are more than likely not the same as for monocultures, measurements in optimal monocultures are not necessarily relevant to the intercrop design. What is needed is an approach that begins with the architecture of individual plants and extrapolates theoretically to the population level. Such an approach is taken in chapter 7 with regard to intercrops involving perennials.

An additional problem that needs to be mentioned is the relative lack of data with respect to light interception in intercrops. Probably because of the technical difficulties mentioned earlier it has not been usual for researchers to make light measurements in intercropping situations. When such measurements are made they do not seem to provide much surprising information. For example, Reddy & Willey (1979) made light measurements throughout the season in a pearl millet–groundnut intercrop and found, not surprisingly, that the intercrop was intermediate between the millet monoculture and the groundnut monoculture with respect to amount of light intercepted.

Partitioning soil resources

That below ground competition is important has been appreciated by plant ecologists, at least since the classic experiments of Watt & Fraser (1933). These workers reduced competition from tree roots by constructing ditches of various depths to surround the understory vegetation, thus reducing any effects of competition through roots. At least two understory species grew

Table 5.1. *Estimated fractions of uptake mechanisms for corn roots* (Donahue *et al.*, 1977.)

	Approximate percentage supplied by		
Nutrient	Mass flow	Root interception	Diffusion
Nitrogen	98.3	1.2	0
Phosphorous	6.3	1.2	90.9
Potassium	20.0	2.3	77.7
Calcium	71.4	28.6	0
Sulfer	95.0	5.0	0
Molybdenum	95.2	4.8	0

better when the effect of tree roots was removed. Subsequently, a number of studies, through an experimental separation of roots and shoots, have unequivocally demonstrated competition through roots (Donald, 1958; Snaydon, 1971, 1979; Martin & Snaydon, 1982; Aspinall, 1960; Gamboa & Vandermeer, 1988). Thus experimental ecologists have verified in no uncertain terms what farmers take for granted, that soil resources are very important.

Soil resources include water and minerals, possibly oxygen in some cases (Greenwood, 1969). Water relations are especially interesting because water is a resource that carries all the other soil resources. It is thus in some sense quite basic (Newman, 1974; Kowal & Andrews, 1973; McCowan, 1973). Some nutrients, such as nitrogen, are readily dissolved in soil water and thus move by means of mass flow, making them just as mobile as water in the soil column, while other nutrients, such as potassium and phosphorous are easily absorbed on the surfaces of soil particles and thus move slowly in the soil (Bray, 1954; Andrews & Newman, 1970; Nye, 1968). Bray (1954) suggested that a useful way of classifying nutrients is in this dichotomous form (mobile versus nonmobile), realizing that any dichotomous classification will eventually break down. In the case of the major plant nutrients of nitrogen, phosphorous, and potassium the dichotomy is really quite good. For example, Table 5.1 gives figures for nutrient supplies to corn roots. As can be seen, almost 99% of the nitrogen is made available through mass flow, while 91% of the phosphorous moves through diffusion, a very slow process. Potassium is somewhat more mobile than phosphorous (in these soils), sulfur and molybdenum are quite mobile, and calcium is in between.

We can make certain tentative deductions about the process of competition from such figures (Trenbath, 1976; Barber, 1962; Barley, 1970). As nutrients

Table 5.2. *Classification of possible mechanisms leading to the competitive production principle (Modified after Snaydon & Harris, 1979.)*

I. Weak competitive interactions
  a. Same limiting resource, but different sources:
    i. partially different times of use (i.e. semiconcurrent crops);
    ii. partially different zones of use (i.e. different root or shoot zonation).
  b. Different limiting resources:
    i. same requirement met by different resource (e.g. N_2 and NO_3 for legumes and nonlegumes);
    ii. different requirements (e.g. light, various mineral nutrients, water).

II. Strong competition (locally).
  a. Environment spatially variable (i.e. mosaic environment).
  b. Environment temporally variable (i.e. seasonal and year-to-year).

are absorbed by roots, an area around the absorbing root, the depletion zone, is formed. This depletion zone, caused by the action of the root in the first place, is thus a region from which no further nutrient can be drawn, at least temporarily. This process is the first of two conceptual processes thought to characterize plant competition for soil nutrients, as described earlier (Goldberg & Werner, 1983; Clements, 1928). It is the effect of the plant on the soil (the second process is, of course, the plant's response to the depleted zone). It is clear that the depletion zones in the case of mobile nutrients will be larger than those of nonmobile nutrients, causing depletion zones of different plants to come into contact with one another generally sooner in the case of mobile nutrients than in the case of non-mobile nutrients. All else being equal, then, it is tempting to predict that competition is not as likely to occur when the limiting soil factor is non-mobile (i.e. shading is more likely to occur before enough phosphorous is absorbed to create overlapping depletion zones), as when it is mobile (i.e. mass flow makes nitrogen available from a large area surrounding the root, rapidly creating overlapping depletion zones). Such has become something of the conventional wisdom in plant ecology (e.g. Harper, 1977, p. 337), although the ubiquitousness of mycorrhizae seems to be altering that view (Janos, 1983; Gunary, 1968; Gerdeman, 1968).

Taking a more empirical approach, Snaydon and Harris (1979) have categorized mechanisms operating in the soil environment that might reduce competition. I have reproduced a modified version of their table in Table 5.2. This would seem to be quite a useful scheme, and, together with the concepts of nutrient mobility, can be rationalized in a very simple theoretical framework. Recall the notion of environmental grain (see Chapters 3 and 4). The two extremes of environmental grain are coarse and fine, coarse when different environments occur in distinct patches, and fine when those same environments occur as diffuse mixtures. In the case of a nonmobile nutrient, it is probably experienced as coarse grained by the plant, that is, it occurs in patches and a plant must be living directly within one of these patches to be able to absorb that nutrient. A mobile nutrient, on the other hand, occurs at an average density over the whole plot, making it irrelevant whether the plant happens to be in a patch or not. Thus a mobile nutrient will tend to show a fine-grained environmental pattern.

Consider a specific artificial example. Suppose that maize requires high quantities of resource A but can be quite competitive against beans if high levels of resource A are actually present, reducing its yield 80% (say). At this high level of resource A the beans are not able to exert any competitive effect against the maize. On the other hand, at low levels of resource A, beans exert a strong competitive effect against maize, reducing its yield 80%. Similarly, maize at this level is too weak to affect the beans competitively. This situation is illustrated in Figure 5.4(*a*). Let us presume the environment is coarse grained, which is to say that in some patches we get 20% of the monocultural yield of maize (the 80% reduction) while in other patches we get 20% of the bean monoculture yield. In both cases the competitively aggressive species does not experience a yield loss due to competition. Assuming that the two patches occur with the same frequency (50% each), we can write

$$\text{intercropped maize yield} = (0.5)\ (0.2)M_c + 0.5M_c = 0.6M_c,$$
$$\text{intercropped bean yield} = (0.5)\ (0.2)M_b = 0.6M_b,$$

where M_c is the monoculture yield of maize and M_b is the monoculture yield of beans. Thus,

$$LER = 0.6M_c/M_c + 0.6M_b/M_b = 1.2,$$

thus demonstrating an intercrop yield advantage.

But what would happen if this environment were fine grained, which in the present context means, what would happen if the nutrient were mobile? In Figure 5.4(*a*) the theoretical graphical relationship is shown for the relationship between the quantity of the resource and the competitive effect. The 'patches' in the environment are the two indicated points, very high and very low nutrients. That is, the environment as we might measure it before the planting season, looks something like the first of the two smaller diagrams in Figure 5.4(*a*), with some patches having much of the resource and other patches having little of that same resource. But since the resource is mobile, it is viewed as a fine-grained environment, which means it is experienced by the plant as a blurred blend of the two patches, as pictured in the second of the two smaller diagrams in Figure 5.4(*a*). As far as the plant is concerned, all of the environment is $0.5x_1 + 0.5x_2$ of the resource, where x_1 is the low concentration and x_2 is the high concentration. From Figure 5.4(*a*) we see that half-way

a.

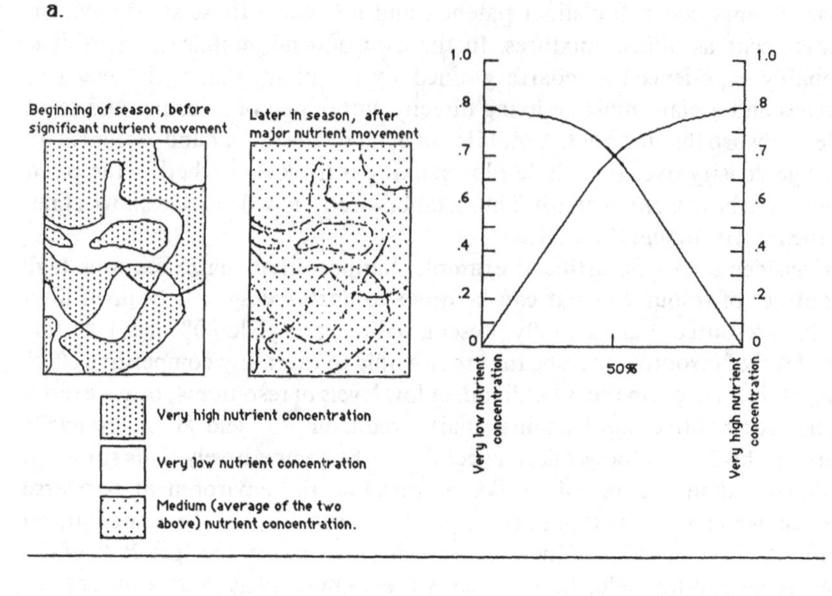

b.

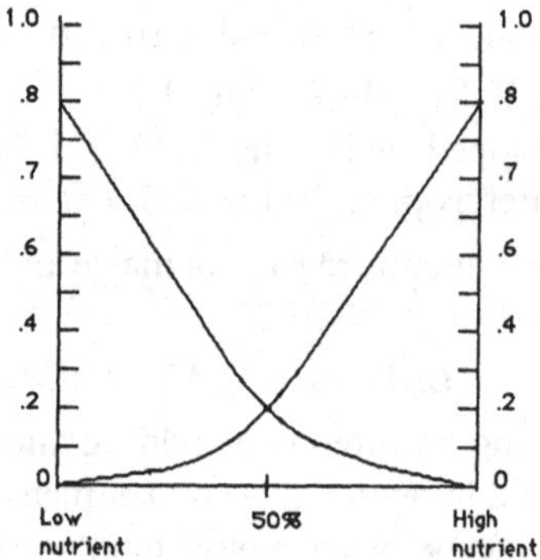

Fig. 5.4. Effect of environmental grain on the process of competition (see text for explanation).

between x_1 and x_2 (because the frequency of each patch is 50%) gives a competitive effect of about 70% (yielding 30% of each species production), so the overall intercrop yield will be

$$LER = 0.30 M_c/M_c + 0.30 M_b/M_b = 0.60.$$

So in this case there would not be an advantage to intercropping, yet the physical environment is exactly the same with respect to quantity. That is, the amount of resource *A* in the environment is exactly the same in both cases, but in one case (coarse grain) the competitive production principle operates while in the other (fine grain) it doesn't.

In this artificial example the numbers were chosen so as to show that environmental grain matters. They could easily have been chosen so as to show that the advantage acrues when the grain is fine rather than coarse, as presented in Figure 5.4(*b*). The point is not that the competitive production principle operates in a coarse- or fine-grained situation, rather that the grain of the environment may dictate, more that the absolute quantity of resource available, whether it does or not. Such a result is perhaps not intuitively obvious. If the coarse-grained situation leads to competitive reduction it corresponds to Snaydon's category II, strong competition. The idea is that within patches competition is strong, but species 1 survives better in one patch, species 2 in the other. But the opposite result is clearly possible, as illustrated in Figure 5.4(*b*). Strong competition within patches might, for simple physiological reasons, be converted to weak competition when patches are averaged.

It is of interest to speculate on the difference between the two forms of competitive response and resource abundance, as pictured in Figure 5.4(*a*) and (*b*) (we note in passing that obviously not all competition–resource curves will be nicely defined as in Figure 5.4 – rather, both curves will frequently be similarly shaped, a fact that does not change the general conclusion). If the competitive effect changes drastically near those resource values that characterize the patches (Figure 5.4(*b*)) the competitive production principle is likely to operate in the fine-grained situation, thus not corresponding to any of Snaydon's categories. If the competitive effect changes slowly near those resource values that characterize the patches (Figure 5.4(*a*)) the competitive production principle is likely to operate in the coarse grain situation, thus corresponding to Snaydon's category II (*a* or *b*). It would thus seem appropriate to alter Snaydon's categories under II (strong competition) to have two subcategories, (1) rapidly changing competitive response and a fine-grained environment, and (2) slowly changing competitive response and a coarse-grained environment. Such results are potentially quite interesting in the case of soil resources, since their occurrence as mobile and non-mobile offers a clear case of fine and coarse grain alternatives. We are thus led to the tentative generalization, apparently in opposition to much conventional wisdom, that competition will be severe as a function of both the environmental grain (which is a function of both underlying environmental pattern and the competitive effect) and the competitive response. It is thus simply not possible to make generalizations such as 'competition will generally be more severe for nitrogen than for phosphorous'.

This theoretical framework awaits empirical justification. The recent ecological work regarding competitive ability along a productivity gradient (Goldberg, 1982; Werner, 1987; Tilman, 1982, 1984; Grime, 1979) does not help much since it does not take into account the pattern of the environment, and plant competition theory at the level of understanding resource dynamics remains at something of a dichotomous level with a great deal of speculation on the one hand and much unguided empiricism on the other (e.g. Harper, 1977).

An exception to this general impression seems to be with regard to nitrogen dynamics with legume and nonlegume mixtures. It is commonly recognized that legume–nonlegume mixtures frequently demonstrate yield advantages over their monocultural alternatives (Trenbath, 1976; Willey, 1979*b*; Agboola & Fayemi 1971; Kurtz *et al.*, 1952; Walker *et al.*, 1954), more so than in combinations of two nonlegumes (van den Bergh, 1968). The usual interpretation of this pattern is that the two species are tapping different nitrogen sources (de Witt *et al.*, 1966; Hall, 1974; Snaydon & Harris, 1979; Sanchez, 1976), the nonlegume from the soil, the legume from the air. It is thus a situation of very low competition, with regard to nitrogen, since the critical resource literally comes from two sources. Most experimental studies seem, at first glance, to confirm this result in that RYT is reduced when nitrogen is added to the system (Martin & Snaydon, 1982; de Witt *et al.*, 1966; Hall, 1974). This interpretation is perhaps partially flawed.

In a soil environment with nitrogen in short supply, any two species will be in competition. If one of the species has another source, as do nodulating legumes, that competitive pressure will be reduced. The idea of adding nitrogen to the system, apparently, is to eliminate it as a limiting factor, thus forcing the two species to compete on some other stage, perhaps another nutrient, perhaps light. If competition is thus changed to a form such that the special characteristics of the legume cannot ensure an independent source of the limiting resource, competition will be as severe in the intercrop as it is in the monocultures and thus RYT will decrease.

The classic experiments of de Witt *et al.*, (1966) conform exactly to these predictions. Working in pots with the grass *Panicum maximum* and the legume *Glycine javanica*, they found relative yield totals on the order of 0.9 when the *Glycine* was not inoculated with *Rhizobium* and no fertilizer was applied. When fertilizer was applied (presumably reducing competition between the two species) the RYT rose to about 1.0. But when *Rhizobium* but not fertilizer was added to the system the RYT rose to about 1.4. Finally, when both *Rhizobium* and nitrogen were added the RYT dropped to about 1.2.

But upon closer examination, the above reasoning seems to fail. If it is true that RYT is expected to increase as competitive pressure is reduced (see Chapters 2 and 3), when a legume-nonlegume combination is subjected to a small increase in nitrogen, the competitive pressure, however small, felt by the nonlegume, will be reduced. Reducing competitive pressure raises RYT. Rather than reducing RYT as a result of nitrogen addition, it is equally

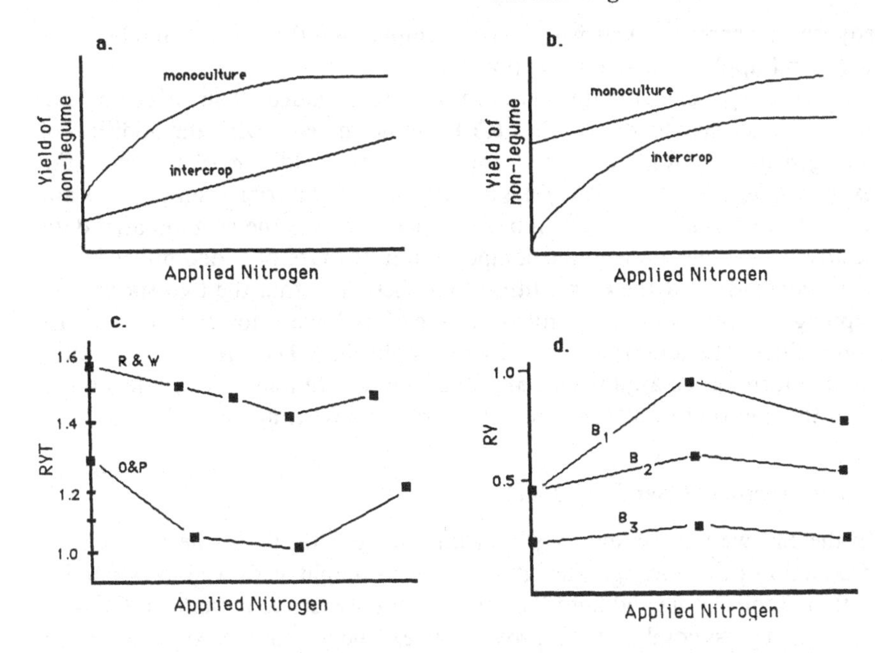

Fig. 5.5. Examples of crop responses to nitrogen augmentation (*a*) the pattern which would show a decrease followed by an increase in RYT; (*c*) exemplary data illustrating the pattern suggested in (*a*) (data from Rao & Willey, 1981 and Oelsligle & Pinchinat, 1975); (*b*) the pattern which would show an increase followed by a decrease in RYT; (*d*) exemplary data illustrating the pattern suggested in *b* (data from Vandermeer, unpublished, tomato–bean intercrop in Michigan).

possible to expect an increase in RYT until the point is reached that nitrogen is saturated in the environment and the focus of competitive interaction shifts to some other factor. In fact, the pattern of change of either the relative yield or relative yield total as a function of added nitrogen simply does not speak to the question of reduced competition. As nitrogen is added to the system both monoculture and intercrop experience reduced competition for nitrogen. It is a question of the rate at which competition is reduced. This concept is illustrated in Figure 5.5. In Figure 5.5(*a*) the reduction in competition is more rapid for monoculture than for intercrop at low nitrogen levels, and thus a pattern of reduced RYT is expected, as shown in data from Oelsligle & Pinchinat (1975; cited in Sanchez, 1976) in Figure 5.5(*b*). In Figure 5.5(*c*) the opposite situation is shown along with exemplary data in Figure 5.5(*d*) (Vandermeer, unpublished). Similarly, Cordero (1977) found a negative relationship between RYT and applied nitrogen when intercropping corn with soybeans, whereas in a corn-snapbean combination the relationship between RYT and applied nitrogen was positive.

Thus, the problem of interpretation is not really aided by any of these data. Since it is as easy to reason that RYT should increase with the addition of nitrogen as that RYT should decrease with the addition of nitrogen, the examination of RYTs (or RYs) as a function of added nitrogen simply does not speak to the issue of competition. As impressive as is the accumulated data relative to legume–nonlegume competition in the face of added nitrogen, we must conclude that the competition hypothesis (i.e. that the two species are tapping different sources of nitrogen and thus have a lower than average competitive interaction) is not really yet established. The data are numerous, but the form of the argument makes them circumstantial. We will deal more with the question of nitrogen dynamics in the next chapter.

The interaction of soil factors and light

In the end we must come to grips with the fact that when the competitive production principle operates, its mechanism might very well be a combination of soil factors and light. The experiment of Stern & Donald (1962) with clover and grass combinations provides an excellent example. At low nitrogen levels the clover, able to tap the atmosphere for its nitrogen, totally dominates the system. At higher nitrogen levels the grass is able to gain more of a foothold in its initial growth, now having a sufficient supply of nitrogen available, but then with the height gained by the applied nitrogen it is able to overtop the clover and outcompete it. Thus at high nitrogen applications the grass dominates the system. What begins as competition for a soil nutrient ends with shading, and it would be absurd to try to understand the system as anything but an interaction of soil factors and light.

A most interesting theoretical treatment extends from the old idea that plants are likely to exhibit a tradeoff between tolerance of low soil resources and shade (or, alternatively, responsiveness to increased soil nutrients and increased solar radiation) (Bourdeau, 1954; Spurr & Barnes, 1980; Tilman, 1985, 1986). A variety of physiological or morphological traits could lead to such a tradeoff, such as differences in root vs shoot allocation and differences in leaf structure (Mooney & Gulmon, 1979; Horn, 1971). In its simplest form the idea is that mixtures of the two different types of species are more likely to satisfy the reduced competition of the competitive production principle.

While a few experimental studies have attempted to disentangle the relationship between light and soil factor competition (Donald, 1958; Snaydon, 1971, Aspinall, 1960; Gamboa & Vandermeer, 1988), the difficult and time-consuming experimental techniques necessary in such studies have not exactly made them popular. The interaction between light and soil factors thus remains at something of a dichotomous level, theory bordering on speculation at one level and empiricism in a theoretical vacuum at the other.

Unlike most purely ecological studies, in intercropping we are concerned with a particularly short time frame, one in which it is difficult to imagine light competition at the beginning (when seedlings begin to grow) and in which it is difficult to imagine a lack of light competition at the end. Generally, we might postulate that during initial growth stages the principle competitive effects are likely to be through soil factors (if competition exists at all), while at the end of the growth period the effects are more likely to be for light. Thus we can conceive of three phases. First is the phase in which seedlings are so small that they do not compete, but nevertheless begin the development of nutrient depletion zones, especially with respect to mobile nutrients (competitive 'effect'). Second is the phase in which nutrient depletion zones are well-established and competitive effects are felt through the reduced soil-factor availability (competitive 'response' to soil-factor depletion). Third is the phase in which above-ground canopies begin to overlap (effect) and photosynthesis is modified according to the reduced light environment (response). Given only these three stages (and this is in and of itself an admitted oversimplification), and the various possible combinations of structurally and physiologically different crop-types, one can see that a great many alternatives are possible for competitive production to take place. A tall-statured plant is likely to exert competitive dominance over a short-statured one, but the latter might very well have been competitively superior in earlier stages when competition was for soil factors. On balance, the two might very well exhibit competitive production, despite the fact that competition is strongly favoring one or the other at any one time. Such is a single example amongst a large number of possible scenarios. That it might never be possible to describe mechanisms at a suitable level of generality suggests that for many practical purposes it will be more appropriate to focus on the phenomenological level, sharpening our theoretical tools here rather than attempting to reduce the problem to nutrients, or physiology, or architecture, or whatever. Such sharpening of theory is exactly what is attempted in a later Chapter (Chapter 10), when we extend the body of theory usually associated with even-aged monocultures to the case of an even-aged diculture.

6

The environments modified to produce facilitation

It seems to be the case, as emphasized in Chapters 3 and 5, that many if not most cases of intercrop advantage are due to the competitive production principle. But many known cases, and probably many that have not yet been sufficiently studied, might very well be due to the modification of some environmental factor, in a positive way, by one of the crops. Certainly, the more spectacular cases must involve some sort of facilitation since competitive production can maximally yield an LER of 2.0, and any individual relative yield can never be greater than 1.0 (see Chapters 2 and 3). In those frequent cases where a relative yield is greater than 1.0, the facilitative production principle must be operative, in some way. This chapter discusses those environmental factors that are typically thought to be modified when facilitation is operative.

Nitrogen

Since many commonly occurring intercrop systems involve a nodulating legume, and since they frequently yield better than their monocultural components (Trenbath, 1976; Snaydon & Harris, 1979; Walker *et al.*, 1954; Allen & Obura, 1983), it is most natural to suspect that nitrogen is somehow involved. Just examining the extensive literature on grass–clover combinations (e.g. Simpson, 1965; Stern & Donald, 1962; Ennik, 1969; Cowling & Lockyer, 1967; Camlin *et al.*, 1983), one is left with little doubt that nitrogen must be of overwhelming importance. But nitrogen can be involved in two distinct ways. First, as already covered in Chapter 5, since each of the species involved is tapping a separate source of a critical resource (the legume N_2 the nonlegume NO_3), it is likely that reduced competition is operative. Second, probably because it has frequently been thought that some special positive mechanism was necessary to account for an intercrop advantage, earlier literature seems to have presumed that the legume added nitrogen to the soil for the system as a whole (Kass, 1978; p. 26). While such an assumption was certainly unwarranted, it nevertheless led to several experiments, most of

Table 6.1. *Experiments involving legumes and nonlegumes*

System	Experimental results	Reference
Oat–soybean	Mixture contained more N than monocultures	Nobbe & Richter, 1902
Barley–pea	Mixture contained more N than monocultures	Pilz, 1911
Maize–cowpea	Mixture contained more N than monocultures	Lipman, 1912
Barley–peas	Non-legume contained more N than monocultures	Kellerman & Wright, 1914
Oat–pea	Mixture contained more N than monocultures	Wartiovaara, 1933
Oat–vetch	Mixture contained more N than monocultures	Nicol, 1934
Barley–peas	Non-legume contained more N than monocultures	Nowotonowna, 1937
Millet–groundnuts	Mixture contained more N than monocultures	Schilling, 1965
Oats–peas	Non-legume contained more N than monoculture	Bodkevitch, 1973
Sorghum–various legumes	Mixture contained more N than monocultures	Wagmare & Singh, 1984
Maize–bean	Mixture contained more N than monocultures	Clark & Francis, 1985

(from Kass, 1978, with modifications.)

which demonstrated the same thing: the legume apparently does transfer nitrogen from the air to the system. A summary of some of these studies is presented in Table 6.1.

Thus at least some of the literature strongly suggests that the role of a legume in an intercrop system is as a source of nitrogen for the system as a whole, or at least for the nonlegume. Actually, the available evidence is only circumstantial, as later experiments with nitrogen addition have clearly shown, as already explained in Chapter 5.

The logic of experiments involving the addition of nitrogen to legume–nonlegume combinations, fully described in Chapter 5, take on special importance when facilitation is involved. While the addition of nitrogen does not imply a predictable response from the system under competitive production, it most definitely does in the case of facilitation. If one crop is actually supplying nitrogen to the other (not just using another source so as to keep out of the way), the addition of nitrogen can reasonably be expected to reduce RYT. Since most studies involving the addition of nitrogen usually report a reduction in RYT, occasionally no response at all, and only rarely an increase in RYT (see Chapter 5), one is tempted to suggest that available evidence supports facilitation rather than competitive production as the operative mechanism in legume–nonlegume combinations. That is, if competitive production were the rule, we would expect some studies to show increased RYTs and others decreased RYTs as a response to N enhancement. But if facilitation is operative, we should always expect a reduction in RYT with N enhancement. Available evidence does seem to support the latter.

But a finer mechanistic question is then generated. If nitrogen is transferred, exactly what is the mechanism of that transfer? It is conceivable that the legume roots are 'leaky' and nitrogenous compounds are excreted into the soil, thereby becoming available for uptake by the nonlegume (Stallings, 1926; Nicol, 1934; Virtanen *et al.*, 1937; Simpson, 1965; Whitehead, 1970). On the other hand, nitrogen may be released as a normal decay process of nodules and roots (Walker *et al.*, 1954; Vincent, 1974; Haynes, 1980). A more active mechanism might involve mycorrhizal fungi in the process (Haystead, 1983). In an elegant experiment, van Kesse and associates were able to demonstrate not only the direct transfer of nitrogen from soybean to maize, but also that the transfer was mediated through vesicular arbuscular mycorrhizae (van Kesse *et al.*, 1985).

In intercroping systems involving nodulating legumes in which the legume is harvested before the nonlegume, if all of the fixed nitrogen is not taken up by the part of the legume that is harvested, and if the legume harvest happens sufficiently prior to the final nitrogen demand by the nonlegume, it is conceivable that the nitrogen tied up in the unharvested parts of the legume could be utilized by the nonlegume.

Such an obvious mechanism suggests a less obvious, but in principle equivalent mechanism, not necessarily restricted to nitrogen. In any system in which a critical source has a flow rate through the system, there is always a possibility that one species might act to reduce the rate of flow of the resource out of the system, with the indirect consequence that another species may benefit (Vandermeer *et al.*, 1986). The classic example is the hermit crab and its shell (Hazlett, 1981, 1983) in which the critical resource is shell. The shell is continually provided to the system by the natural death of mollusks, but is also continually removed from the system by gravity and wave action, unless it is occupied. Thus if one species occupies shells that would normally fall out of the system, such that another species can eventually use them, the first species has an indirect facilitative effect on the second (Hazlett, 1981). This principle is easily generalizable to any resource that flows into and out of a system. Nitrogen is a particularly good candidate since its rate of movement through the soil is so rapid. If one species (the legume, perhaps) absorbs quantities of nitrogen that would otherwise leach out of the system, and some of that absorbed nitrogen is later made available to the nonlegume, facilitation will occur. Exactly this mechanism was observed by Agamuthu & Broughton (1985) working on oil palms. In a young palm plantation the trees are small and dispersed, and their relatively underdeveloped root mass cannot utilize available nitrogen from the extensive areas between individual trees. An interplanted legume not only fixes atmospheric nitrogen, but also scavenges nitrogen from the soil areas that the palms are unable to reach, thus effectively 'banking' the nitrogen in the legume plants. When the oil palm roots grow into the area occupied by the legumes, the latter act to slowly release the accumulated nitrogen to the oil palm roots.

Since luxury consumption is apparently a common feature of many crop species, the above mechanism seems likely to occur. Even if it doesn't occur routinely, it is nevertheless a candidate for intercrop improvement. In a locality in which nitrogen moves very rapidly through the soil (in tropical oxisols and ultisols, for example) it may be possible to use early luxury consumption as a sort of storage for applied nitrogen. Instead of rapidly leaching out of the system, a cereal intercrop (say) could be used to trap the applied nitrogen and later release it to the desired crop.

As to whether or not nitrogen is involved in legume–nonlegume intercrops there seems to be little doubt. Whether it is simply a case of one crop using one source (NO_3) and the other using another (N_2), a simple application of the competitive production principle, or whether the legume is actually transferring nitrogen into the system, which would be an application of the facilitative production principle, might seem like an overly academic exercise. But I think not. In many situations, especially in the Third World, but also amongst the more impoverished farmers of the developed world, nitrogen fertilizer is an important and expensive input. If it is true that the legume is actually supplying nitrogen to the system, it becomes of potential practical interest to such sectors. It may be quite practical to initiate an intercrop of a cereal and legume, only for the nitrogen input the legume may provide, whether or not the legume is actually harvested.

Water

It is widely recognized that competition for water can be an important component in any combination of plants growing together. Its discussion under the topic of the competitive production principle was thus perfectly obvious. But only rarely has the obvious fact that a second species can positively modify the water environment been mentioned. The obvious cases are with 'cover' crops such as clover (Simpson, 1965; Stern & Donald, 1962; Walker *et al.*, 1954; Ennik, 1969) and with shelter belts in agroforestry (Leyton, 1983).

The addition of a second crop to some monoculture has a potential for augmenting the water environment through the reduction in evaporation, in two ways. First, if the soil is covered with vegetation, as it is in the case of a cover crop, evaporation is likely to be reduced. Second, if a windbreak is created, the input of advected energy is lowered, thus lowering evaporation. Either or both cases could operate in intercropping systems.

As already discussed in the case of nitrogen, the experimental addition of water to an intercrop does not provide clearcut predictions with regard to changing RYTs. If competitive production is the case, the RYT will either increase or decrease with addition of water, while it can only decrease in the case of facilitation.

In a study of a tomato cucumber intercrop in Michigan, Schultz (1984) found increasing RYTs as a function of irrigation. Other examples (Kurtz *et al.*, 1952; Fisher, 1977; Singh et al., 1973; Lenga & Stewart, 1982; Shackel & Hall, 1984) have given equivocal results, sometimes showing improved intercrop performance with added water, sometimes no effect, and sometimes reduced intercrop performance.

One situation in which facilitation is clearly implicated is in the use of a secondary species as a shelterbelt, either with annual systems (Radke & Hagstrom, 1976) or perennial shelters (Leyton, 1983). Radke & Hagstrom discuss the wind-breaking aspects of intercropping, reviewing examples including corn sheltering soybeans (Radke & Burrows, 1970), sunflowers for soybeans (Radke & Hagstrom, 1974), corn for sugar beets (Rosenberg, 1966), and oats sheltering tomatoes (Short & Kretchman, 1974). A large number of studies using snowfences or slat fences are also cited by Radke & Hagstrom. These studies are interesting from a biological point of view because they impose experimentally the environmental factor that the intercrop supposedly modifies, without adding the full competitive effect of the secondary species.

For example, Rosenberg (1966) compared snowfence windbreaks with corn windbreaks in protecting irrigated sugarbeets in Nebraska, U.S. The snowfence produced a 15.2% increase in sugarbeet yields while the corn strips produced a 26% increase. In 15 cases of corn barriers for soybean, percentage increases in soybean production ranged from −5 to 28%, while two cases of snowfence barriers produced 0 and 10% advantages (Radke & Hagstrom, 1976). From other experiments they report that corn formed the most effective barrier, followed by sunflower, followed by snowfence and, finally, a solid board fence. Rosenberg & Hagstrom conclude 'crops sheltered by temporary wind barriers tend to grow taller, produce more dry matter, with a larger leaf area index and larger yields. Wind barriers alter the microclimate which in turn alters the plant water relationships of the sheltered crop.'

Leyton (1983) with an emphasis on agroforestry, discusses the problems inherent in providing a shelterbelt and the competitive pressure imposed by a living shelterbelt. The shelterbelt partially shades the crop, which is both good and bad – maintaining moisture through shade, but reducing potential photosynthesis. The shelterbelt reduces evaporation by decreasing wind, but also increases demand on soil nutrients.

In general, the technique of shelterbelts, whether trees or annual crops, fits in nicely with the analytical framework offered by the facilitative production principle. The secondary crop is the shelterbelt, and its scale is the relative number of strips per unit area. In a theoretically wind-free environment, the only effect of the shelterbelt would be negative. In a maximally windy environment, the net effect would be positive.

That water as a resource is perfectly compatible with the formulation of the facilitative production principle, as discussed in Chapter 4, is made clear in figure 6.1. Figure 6.1(*a*) presents the simple relationships between degree of intercrop and amount of water available to the principle crop (the amount of water eventually translating into yield, thus corresponding to the introductory graph of Chapter 4). Figure 6.1(*b*) illustrates what the optimal intercropping strategy would look like.

Non-nitrogen nutrients

Scant attention has been paid to the role of nutrients other than nitrogen in intercropping research. Yet from the small amount of literature available, the results are quite suggestive (Kass, 1978). For example, it has been shown several times that there is more P in the intercrop than in either monoculture (Dalal, 1974; Agboola & Fayemi, 1971; Kaurov & Budkevitch, 1973). Similar evidence is available for K, Ca, and Mg, although in all cases the evidence is scanty. Generally, according to Kass (1978): 'It must be concluded from the small number of experiments in which measurements were taken that crop mixtures will contain greater amounts of P, K, Ca and Mg than will pure stands of the component crops grown under the same conditions.'

However, as discussed earlier, this type of evidence does not really speak directly to the question of facilitation. Again, as stated by Kass: 'The available evidence indicates that crop mixtures take up more applied and soil P than do monocultures of the same crops.' The implication here is quite interesting. It could be that there really are two different sources of these nutrients (from roots at different levels in the soil, for example) and the increase in total nutrient taken up by the combination is nothing more than a reflection of competitive production. On the other hand, it could be that one of the crops is actually mining a source not available to the first and somehow making it available, implying the operation of facilitation. It would seem that this is a whole area of research that awaits some detailed experimentation.

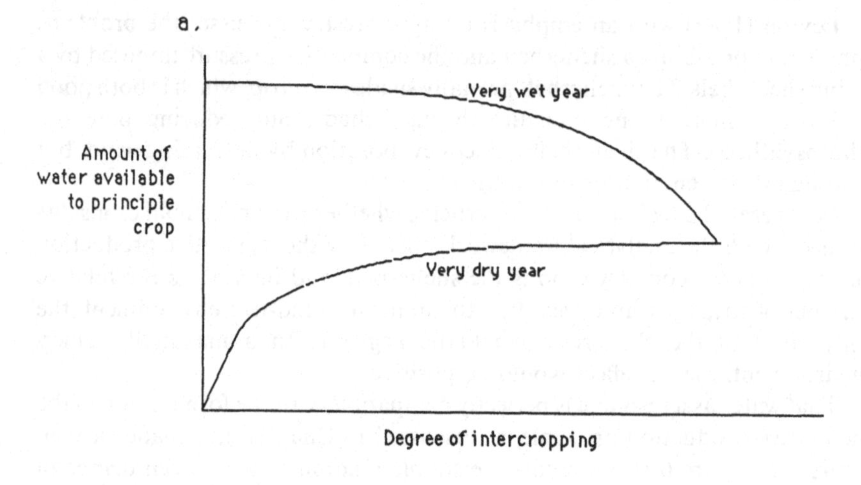

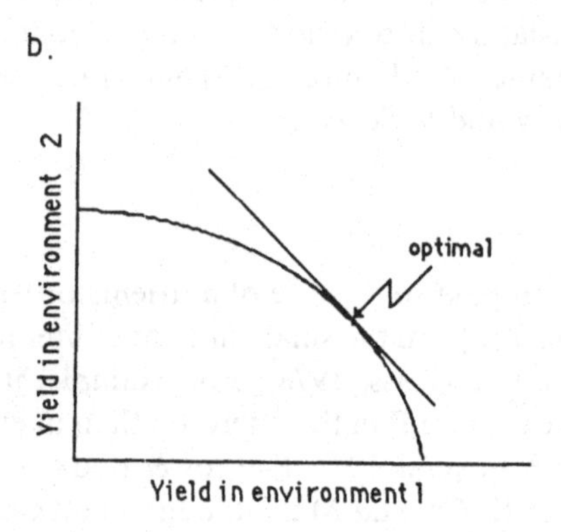

Fig. 6.1. Application of the facilitative production principle to the question of water use of an intercrop. Note similarity of the formulation to Figs. 4.9–4.12 (see text for explanation).

One factor that has been relatively ignored until recently is the relative ubiquitousness of mycorrhizal fungi. 'Harley considered that in some cases the root was so unimportant as to be an artifact. This is because practically all plants in the natural condition are mycorrhizic and it is through mycorrhizae that nutrient uptake from soil occurs.' (Gunary, 1968). Indeed, the striking results of Chiariello *et al.* (1982), in which it was demonstrated that *P* was actually transferred from one species to another through mycorrhizal connections, is of critical importance here (as was the case with nitrogen transfer described earlier (van Kesse *et al.*, 1985)). If mycorrhizal connections are regularly formed between species, and one species can mine different sources of nutrients than another, the ability to transfer the otherwise unavailable nutrients represents an obvious facilitation. Possibly this was the operative mechanism in the experiments of Cecilio & Janos (1987), in which the intercrop advantage of a maize–solanum intercrop was lost in a mycorrhizae-free environment. It might even be the case that most situations in which partititioning of the soil environment would normally lead us to think of reduced competition actually represent facilitation. Mining a different part of the soil column is one thing, but transferring what was mined to another species which would normally not be able to obtain it, is quite another.

Protection from pests

Of the variety of factors that might be involved in the facilitative production principle, the one most cited and perhaps best-documented is the reduction in pest attack frequently found in intercrops. Table 6.2 presents a review of the literature on this subject. Earlier reviews found similar results (e.g. Perrin, 1977; Kass, 1978; Nickel, 1973; Litsinger & Moody, 1976; Dempster & Coaker, 1974), that pests tend to be reduced in intercrops, although not by any means always. While these reviews tend to concentrate on insects, there is also evidence that intercrops reduce nematode attack (McBeth & Taylor, 1944; Khan *et al.*, 1971; Atwal & Mangar, 1969; Castillo *et al.*, 1976; Egunjobi, 1984) and diseases (Moreno & Mora, 1984; Rheenen *et al.*, 1981).

It is thus clear that diversified agroecosystems, of which intercrops are the prime example, frequently show reduced herbivore attack, although the pattern is highly variable. Indeed, for the same cropping system and the same insects, results can vary. Francis and coworkers found lower attack rates of *Spodoptera frugiperda* in a maize-bean intercrop as compared to a maize monoculture (Francis *et al.*, 1978). Van Huis (1981) working in Nicaragua found the same pattern with the same pest in the same cropping system. But my own work on this system in Nicaragua has repeatedly failed to detect such a difference both under rainfed and irrigated conditions (Vandermeer, unpublished data).

The pattern of reduced herbivore load in intercrops apparently results from a variety of causes. Making categorical sense of these varied causes has occupied considerable space in the literature. Trenbath (1976) proposed a self-explanatory metaphor, the 'fly-paper' effect, in which a specialist pest is deterred from its host through the disruptive effect of an associated species of plant. Aiyer (1949) formulated a three-part hypothesis, to wit: (1) host plants are more widely spread in intercrops, meaning they are harder to find, (2) one species serves as a trap-crop to detour the pest from finding the other crop, and (3) one species serves as a repellent to the pest. Root (1973) suggested a pair of

Table 6.2. *Numbers of monophagous and polyphagous herbivore species more or less abundant in diversified agroecosystems compared with monocultures for annual and perennial cropping systems. Numbers in parentheses are percentages*

System	More abundant	No difference	Less abundant	Varied	Total
Annual					
monophogous	3	15	58	23	99
polyphagous	16	2	11	12	41
total annual	19(14)	17(12)	69(49)	35(25)	140
Perennial					
monophagous	12	1	34	4	51
polyphagous	5	0	2	0	7
total perennial	17(29)	1(29)	36(62)	4(7)	58

(Risch *et al.*, 1983.)

alternative hypotheses, the 'resource concentration hypothesis' and the 'enemies' hypothesis, in which the intercropping situation either provided host plants in a nonconcentrated fashion (similar to a combination of 1 and 3 of Aiyer), or acted to attract predators (enemies) more than in a monoculture, thus reducing the pests in the intercrop. A large amount of excellent experimental work was stimulated by the proposal of Root (e.g. Risch, 1980, 1981; Bach, 1980*a*, 1980*b*, 1981; Hansen 1983; Cromartie, 1980). While some degree of advocacy seems to have evolved concerning these alternative hypotheses, it is well to remember that all have been convincingly demonstrated in one system or another, and it is not, nor was it ever, a question of one being generally right and others wrong.

None of the proposed formulations really includes all the mechanisms that are known to operate in the general area of diversity and pest attack. Aiyer's categories do not include predators and Root's do not include trap cropping (at least explicitly). It seems worth proposing an alternative scheme, one which encompasses all the schemes proposed thus far. Let us conceive of the mechanisms leading to reduced pest attack in intercrops as being divisible into three categories.

1 the disruptive-crop hypothesis (part of Trenbath's fly-paper effect and Root's resource concentration hypothesis), in which a second species disrupts the ability of a pest to efficiently attack its proper host, largely applicable to specialist herbivores;

2 the trap-crop hypothesis (a second aspect of the fly-paper effect), in which a second species attracts a pest that would normally be detrimental to the principle species, largely applicable to generalist herbivores;
3 the enemies hypothesis, in which the intercropping situation attracts, for whatever reason, more predators and parasites than the monocultures, thus reducing the pests through predation or parasitism.

Disruptive crop is equivalent to 1 and 3 of Aiyer while trap crop is equivalent to number 2 of Aiyer. Disruptive and trap crop together are the flypaper effect of Trenbath and perhaps equivalent to the resource concentration hypothesis of Root. Bach (pers. com.) prefers the dichotomy resource concentration, referring to any mechanism involving the interaction of the pest with its host plant, versus enemies, referring to any mechanism involving the next trophic level. The classification used in this book is consistent with Bach's in that disruptive crop and trap crop taken together become the resource concentration hypothesis. I insist on this separation for practical convenience – they each require unique theoretical treatments and imply distinctly different strategies for intercrop management. We consider these three hypotheses in turn.

The disruptive crop hypothesis

The general behavior of ovipositing tomato hornworms (*Manduca quinquimaculata*), is a picture of the operation of the disruptive crop hypothesis (personal observations). When a female approaches a clump of vegetation she touches the vegetation with her front tarsi, presumably obtaining a chemical cue from the plant. If it is a tomato that she encounters she immediately lays an egg, followed by a quick exit to hover about 10–20 cm above the plant. Another encounter with the plant results in another egg being laid; this process is repeated 5–10 times, before she flies off. On the other hand, if the clump of vegetation is a tomato surrounded by beans, the moth frequently encounters a bean leaf rather than a tomato leaf, resulting in a quick exit to hovering position, but without laying an egg. There are 5–10 encounters regardless of the composition of the vegetation, before flying off. But if the wrong host plant is encountered, no egg laying takes place. The result is generally more oviposition in tomato monocultures than in tomato–bean intercrops (on the other hand, perhaps because of predation, the actual number of larvae found in intercrops is not much smaller than in monocultures).

More generally, the disruptive crop hypothesis may operate when a herbivore (*a*) is less likely to find its host plant because of some kind of confusion (chemical or physical) imposed by a second species, and/or (*b*) after having found a host plant patch, is more likely to leave that patch because of encounters with nonhost plant individuals (Bach, 1980*b*). The idea is closely related to the notion of plant apparency (Feeny, 1976), in which some factor causes individual host plants to be less 'apparent' to mauraudeing herbivores. In the case of intercrops, the primary crop is made less apparent to pests through some action of the secondary crop. This hypothesis might also be true for less obvious reasons. The disruptive species may exert its influence by creating a microenvironment which is less suitable for the pest, or which leads to lowered reproduction of the pest.

Perhaps the most classic, if somewhat trivial, example of the operation of the disruptive crop hypothesis is the dense rows of thorny plants which exclude vertebrate pests in India (Aiyer, 1949). Working with a corn–bean–squash system in Costa Rica, Risch (1980, 1981) gathered much evidence in support of the disruptive crop hypothesis for several species of beetle pests on both corn and beans. Bach (1980*a*, *b*) collected similar evidence for the cucumber beetle (*Acalyma vittatum*), with the additional feature that she eliminated the simple question of density from the system (i.e. distinguished Aiyers' alternatives 1 and 2 see above), one of the few reported studies to have done so. Rosset (1986) interpreted the reduction of *Heliothis* sp. and *Liriomyza* sp. on tomatoes to be a result of the interfering effect of intercropped beans. Working with the collard pest *Phyllotreta cruciferae*, Cromartie (1975) found that the vegetational background, analogous to the secondary crop, had a significant effect on the colonization of crucifers by their herbivores. Bergelson & Kareiva (1987) demonstrated that imposing a curtain barrier between plots of monocultures and polycultures, and thus generally reducing the mobility of the herbivore, reduced the effect of intercropped potatoes on *Brassicae oleraceae.* Similarly, when an intercrop breaks up the normally sharp distinction between host plant and soil, aphids are less efficient at locating their hosts (Smith, 1969; Halbert & Irwin, 1981). (For a more extensive review of the general effects of plant patterns on the movement of insects, see Stanton, 1983 or Kareiva, 1983.)

In all cases mentioned, the evidence for the disruptive effect of a second species, however convincing, is indirect, and is only rarely tested as a true alternative to the enemies hypothesis. While the evidence in favor of the operation of the disruptive crop hypothesis is certainly available, it is not usually the form of excluding other hypotheses, specifically the enemies hypothesis, other than a simple failure to detect more 'enemies' in the intercrop. Also, the joint operation of the two hypotheses is a clear possibility. If a generalist herbivore, having been diverted from a nearby hostplant because of the confusing effect of the chemistry of a neighboring intercrop is suddenly attacked by one of the numerous predators that had been drawn into the system by the nectar offered by the flowers of that same intercrop, which of the two alternatives is actually operating? The work of McGuinness (1987) directly attacked this question. Working with the Mexican bean beetle, *Epilachna varivestis*, in a tomato–bean intercrop, he constructed detailed life-tables for both monoculture and intercrop. Using the life table data he was able to show convincingly that both the disruptive crop hypothesis and the enemies hypothesis were operative, and he further developed a quantitative technique for estimating the importance of each.

The disruptive crop hypothesis might also be true for indirect reasons. In an elegant experiment, Bach (1981) reasoned that plant 'quality' might be affected by intercropping such that the individual host plants in intercrops may be less desirable to their pests than individuals in monocultures. Bach found that *Acalymma vittatum* preferred cucumber leaves taken from monocultures to those taken from cucumber plants intercropped with tomatoes. Thus it might be postulated that the disruptive crop hypothesis works indirectly, the secondary species actually altering the quality of the primary species, from the point of view of the pest.

The disruptive crop hypothesis is equally applicable to the incidence of diseases as to the attack of herbivores, although the experimental evidence is not as voluminous. It seems to be generally accepted that a secondary crop may reduce the dispersability of disease organisms (Leonard, 1969). For example, the bacterial speck (*Pseudomonas tomato*) is an airborne tomato disease that is aided significantly by the presence of small lesions in the plant epidermis. Such lesions may be less common in intercrops if the secondary species is effective in providing a wind-break (e.g. Radke & Hagstrom, 1976). There is some evidence that this mechanism is operative in tomato–rye grass intercrops in Michigan (Price, per. com.). A further interesting aspect of this disease is its apparent effect on the herbivore *Manduca sexta*, the tomato hornworm. Tomato leaves with extensive symptomatology of *P. tomato* infection apparently contain a toxin which kills *M. sexta* (Hansen, pers. com.). This observation sets up an interesting paradox, in which the disruptive crop hypothesis might operate at two levels, but in opposite directions. Lower oviposition behavior in intercrops (as described above) could be offset by greater larval survivorship due to the secondary crop's control of the bacterial speck disease.

Dissecting the operation of disruptive crops a bit further, the work of Kareiva (1982) is relevant. We consider an intercrop as a three-compartment system, host plant (primary crop), alternative plant (secondary crop), and 'out of system'. The probabilities associated with transitions among these three stages generally define a Markov process, which can be used to gain insight into the operation of the process. Specifically, let P_{ij} represent the probability that an insect currently located in compartment i will be found in compartment j one time unit from now. The process is illustrated in Figure 6.2. Mathematically, we have the matrix equation

$$\begin{vmatrix} P_{11} & P_{12} & P_{13} \\ P_{21} & P_{22} & P_{23} \\ P_{31} & P_{32} & P_{33} \end{vmatrix} \begin{vmatrix} N_1(t) \\ N_2(t) \\ N_3(t) \end{vmatrix} = \begin{vmatrix} N_1(t+1) \\ N_2(t+1) \\ N_3(t+1) \end{vmatrix}$$

Where $N_i(t)$ represents the number of individuals of the herbivore in the ith compartment at time t. This equation can be written in matrix form as,

$$\mathbf{Tn}(t) = \mathbf{n}(t+1),$$

where **T** is the matrix of the transition probabilities, and **n** is the vector of the number of herbivores per compartment at time t and/or $t+1$. If the probabilities are constant, the system will eventually reach equilibrium, at which point

$$\mathbf{Tu} = \mathbf{u},$$

where **u** is the eigenvector and represents the equilibrium number of individuals in each compartment. Each element of **u** can be directly calculated from the individual probabilities, which is to say, if we can know what the transition probabilities are, we can calculate how many individuals (or at least what proportion of the population) will be in each compartment.

This approach is interesting for two reasons. First, one can easily examine how a change in one of the probabilities will affect the final population density in each of the compartments, making it possible to predict, from movement patterns, what will be the ultimate proportions of the pest population that will be found in (or will encounter) the host plant versus the secondary plant. Second, the sorts of design opportunities the planner or producer might envision are intuitively related to the probabilities. Thus, for example, increasing the relative concentration of the secondary crop is certain to decrease the value of p_{13} (the probability of transferring from 'outside the system' to the host plant), since many migrating individuals are more likely than before to be disrupted by the higher density of the secondary crop.

Lawrence's (1987) detailed experiments are interesting in this light. The flight behavior of the beetle *Tetraopes tetraophthalmus* can be divided into three qualitatively distinct forms, turns in alternate directions, turns in the same direction, and turns in random directions. Depending on sex ratios, within patches of appropriate host plant the pattern consists mainly of turns in the same direction, while outside of the patch the turns are in alternate directions. Such a pattern suggests that if a secondary crop keys in an 'out of patch' behavior, however slight, the tendency will be to leave intercrops faster than monocultures. Or, in terms of the Markov approach, P_{11} is larger in monoculture than in polyculture.

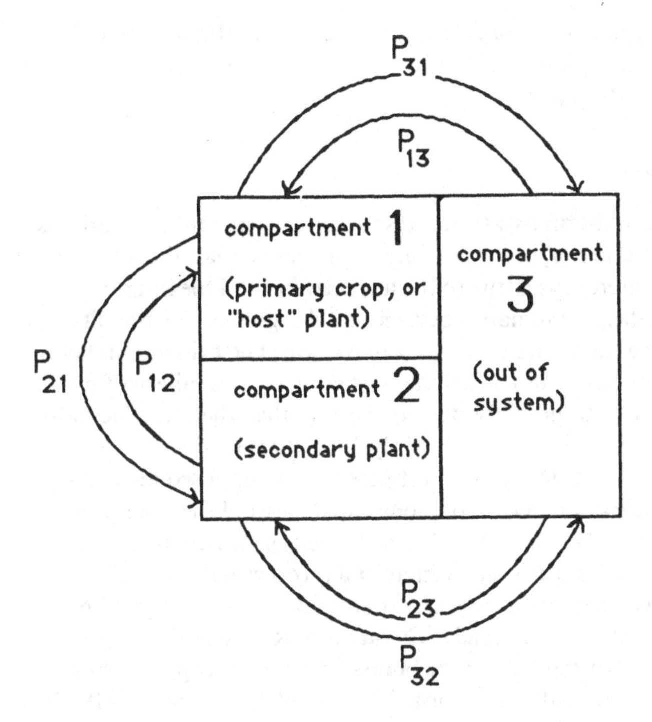

Fig. 6.2. Diagrammatic representation of Markov model for insect movements into and out of an intercropping system P_{ij} represents the transition probability from compartment j to compartment i.

To the determined mechanist and/or reductionist the Markov approach may seem hopelessly phenomenological. That is, by measuring movement patterns and using those movement patterns to describe distribution patterns, one is not getting at the *mechanism* of the movements in the first place. While such a critique can, of course, lead to rampant eclecticism (each type of host plant, secondary plant, surrounding environment, and herbivore have such specific characteristics that a special theory must be developed for every combination), there nevertheless does seem to be an approach which is perhaps closer to 'mechanism' than the Markov process approach, the use of diffusion equations.

This relatively new approach involves partial differential equations and diffusion dynamics (Kareiva 1986). The main idea is that the trivial movements (movements within a field as opposed to movement to or from the field, Southwood, 1978) of herbivores in a monoculture or polyculture can be viewed as a diffusion process, much like molecules diffusing through a space.

This technique can be applied to specialist or generalist herbivores as well as natural enemies, and is explored further in Chapter 11, when the question of planning intercrops is discussed.

The enemies hypothesis

Posed originally as an alternative to the resource concentration hypothesis (Root, 1973), the enemies hypothesis simply proposes that the observed reduction of pests in intercrops is due to the attractiveness of the intercrop for predators and parasites, presumably because of the greater availability of habitats or resources as compared to the monoculture. The evidence in favor of the enemies hypothesis is not nearly as strong as that available for the disruptive crop hypothesis, possibly due to the fact that the latter actually operates more commonly in nature, or possibly due to the fact that the enemies hypothesis has not been as vigorously investigated as its supposed alternative.

There are, nevertheless, numerous demonstrated cases of the operation of the enemies hypothesis. Hansen (1983) clearly demonstrated increases in abundances of several predator species in an intercrop system of maize and cowpea in southern Mexico, suggesting an explanation for the overyielding of that system (Vandermeer *et al.*, 1983). Gavarra & Raros (1975) reported spiders to be more effective against corn borers in an intercrop of corn and groundnuts than in a monoculture of corn. Altieri *et al.* (1977), Smith (1969), and Speight & Lawton, (1976) reported a higher incidence of predators in a weedy crop than in a comparable 'monoculture'. Crookson (1976) demonstrated that grain sorghum attracts predators which can effectively control cotton pests. Perfecto *et al.* (1986) demonstrated that carabid beetles emigrated more rapidly from patches of monocultures of tomatoes and beans than from intercrops of the two.

The theoretical possibilities associated with the enemies hypothesis are much the same as in the case of the resource concentration hypothesis, with the important exception that it is the natural enemies that are being modeled, not the herbivores. Thus, the movement of predators through the system may be regarded as a Markov process or as a problem of diffusion. The interaction of predators and their herbivores, who it must be assumed are also somehow responding to the intercrops, makes the problem intrinsically more difficult in the case of natural enemies.

Trap-cropping

The final mechanism in which an intercrop advantage acrues from the reduction of pests is in the use of trap-crops (Aiyer, 1949; Stetner, 1976, cited in Lamberts, 1980). None of the traditional reviews of intercropping include this

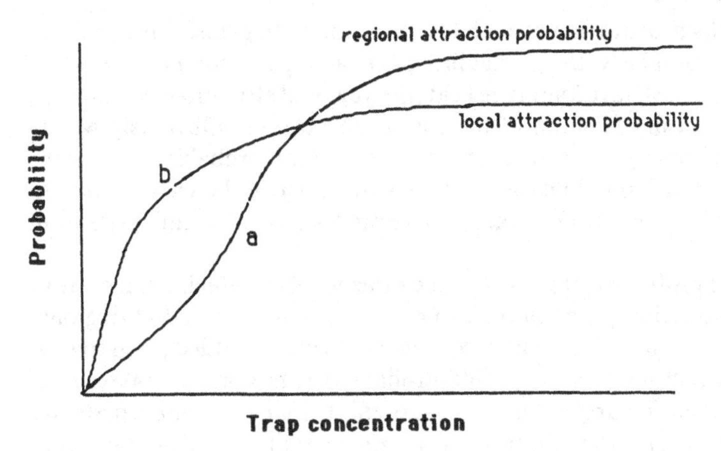

Fig. 6.3. Likely pattern of attraction probabilities for a trap-crop.

mechanism (e.g. Kass, 1978; Willey 1979*a*, 1979*b*; Litsinger & Moody, 1976), even though it is frequently a most obvious mechanism.

The idea is that the presence of a second crop in the vicinity of a principal crop attracts a pest which would otherwise normally attack the principle crop. As early as 1944, McBeth & Taylor reported on the ability of certain cover crops to attract root knot nematodes and thus lower their incidence on peach trees. Corn, when planted in strips in cotton fields, reportedly may attract the cotton bollworm away from the cotton (Lincoln & Isley, 1947), although not always (Bebbington & Allan, cited in Bhatnagar & Davies, 1979). Sorghum is an effective trap for the stem borer, *Chilo partellus* in India (Sarup *et al.*, 1977). The data of Rosset *et al.* (1986) are interesting in this context. An unexpected attack by the army worm (*Spodoptera sunia*) totally destroyed a monoculture of tomatoes, while an intercrop of tomatoes and beans was effective in reducing the attack to virtually zero. It was clear that the caterpillars were being attracted to the beans in the intercrop. By definition, the beans were a trap-crop.

The concept of a trap-crop may be conveniently contrasted to the disruptive crop hypothesis. The latter is thought to act in some way to disrupt the ability of a specialist herbivore to find and remain on its host plant. Trap-crops act to attract generalist herbivores in such a way that the plant to be protected is not as likely to be directly attacked. And here lies a fundamental contradiction. The idea is to have a system in which the trap-crop is 'concentrated' enough to attract the pest away from the crop, but not so concentrated as to attract more pests into the general area. Figure 6.3 illustrates a very simple model demonstrating these two countervailing tendencies. The probability of attracting an individual pest away from the crop after it has already arrived in the field (the 'local attraction probability') is an increasing function of the trap concentration. Similarly, the probability of attracting an individual pest that otherwise would not have found the field (the 'regional attraction probability') is also an increasing function of the trap concentration. Obviously we are concerned with finding cases where the local attraction probability minus the regional attraction probability is positive, that is, where the trap's ability to draw pests away from the crop exceeds its drawback of attracting pests from afar.

The central problem is that exactly the same forces producing the positive effect (local attraction probability) also produce the negative effect (regional attraction probability). Each unit of concentration of trap added (each row, or each plant, or each strip) adds to both probabilities. In very general terms, if the regional effect is larger than the local effect, that is, if more pests are attracted from outside the field by the traps than are drawn to the traps from within the field, the trap-crop will not function. It is relatively easy to visualize this process graphically. Let N symbolize the population density of the pest throughout the whole region (the 'global' population density), a represent the fraction of that population that arrives to the field in question over some specified time period, and b be the fraction of the local population that is attracted to the trap crop. Then if n is the population density of the pest in the field (the 'local' population density), we have, approximately,

$$n = (a/b)N.$$

The two rates, a and b, more or less correspond to the two probabilities of Figure 6.3. Rather than examining these rates separately, looking directly at their ratio we can, first, intuitively picture what they must look like as functions of trap concentrations and, second, directly stipulate if the trap-crop is possible and, if so, exactly what trap concentration is relevant. This procedure is laid out in Figure 6.4.

Using Figure 6.4 we can visualize which trap concentrations will be effective and, moreover, which trap concentrations will be optimum. A further observation, of much practical importance, is that it is quite possible to have trap densities that will not work, even though the secondary crop could be an effective trap. A single experiment to 'try out' a trap-crop could fail, even though the proposed species might be potentially effective at its optimum concentration. Rather than pursue brute force empiricism, it might be more prudent to attempt to estimate the rates directly, in which case the whole curve could be constructed and the optimal trap concentrations directly computed.

A final subject associated with trap-crops is often not fully appreciated. Sometimes for a trap to be effective, it will probably be necessary to plant the trap-crop in fairly high concentrations, leading to two problems. First, the secondary species (in this case the trap) may very well exert competitive

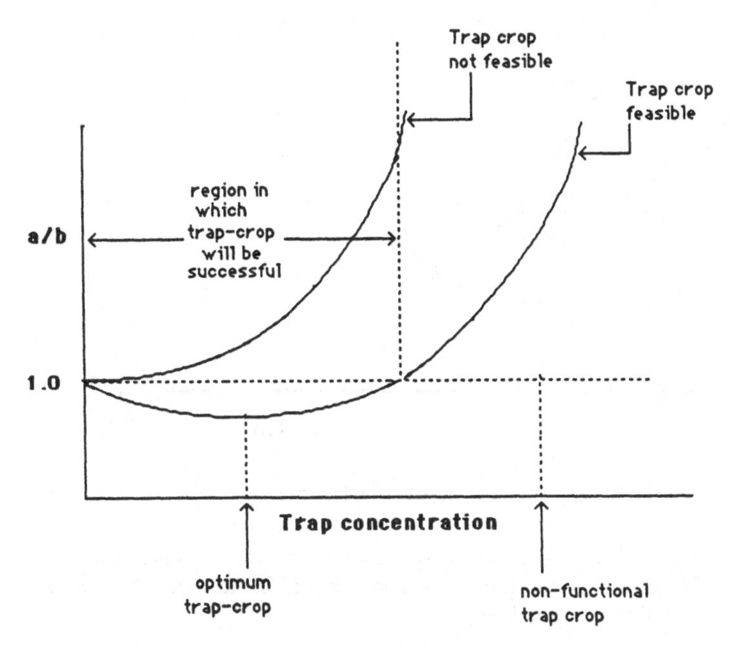

Fig. 6.4. Ratio of general to local attraction probabilities as a function of trap concentration, illustrating the region of trap-crop feasibility.

pressure on the principle crop. Furthermore, in the specific case of trap crops we assume that the trap itself is of no economic importance, which means that whatever area is taken up with trap is lost production of the crop of interest. Thus both the competition effect and the space effect must ultimately be taken into account in the final design of the trap-cropping system. If we formulate the postive effect of the trap as above (a combination of regional and local movement patterns which lead to a net benefit, at least at some concentrations of the trap), we can then contrast it to the negative effects of competition and space preemption, as was done in the simple example of Chapter 4. Alternatively, it is possible to be more analytical if information relevant to the competitive effect of the trap is available, a subject laid out in considerable detail in Chapter 11. Suffice it to say here that with some rather cumbersome equations it is quite possible to balance the positive and negative effects of trap-cropping to obtain an optimum placement of the traps.

The detection of facilitation

At this point a note of caution concerning methodology is in order. The usual agronomist's methods of plot experiments, designed to be microcosms of

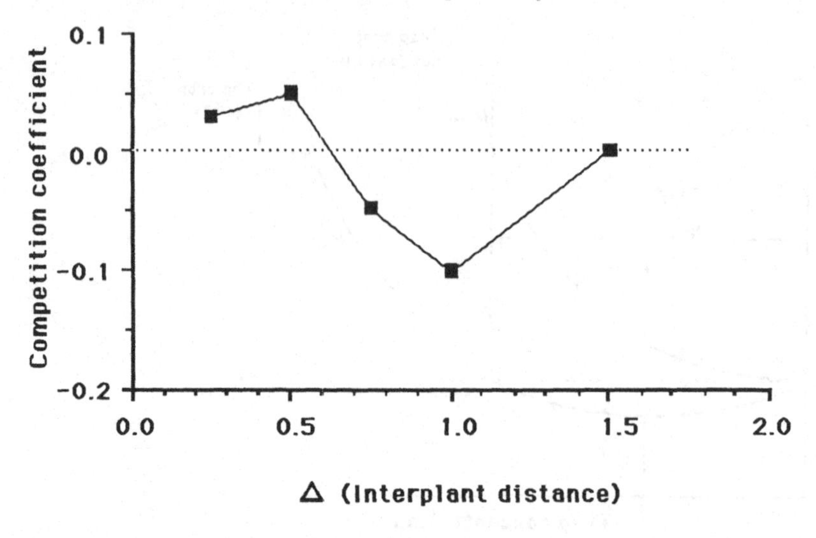

Fig. 6.5. Interaction coefficient (positive values are competitive, negative values are facilitative) as a function of distance to secondary species for the effect of tomatoes on beans (data from Vandermeer, 1986).

actual production, can be misleading. If an experimental plot yields a relative yield greater than 1.0, it is certainly true that facilitation has been demonstrated. But the reverse is not true. Failure to obtain a relative yield greater than 1.0 does not imply the absence of facilitation. Since we can normally expect competition to be operative simultaneously with facilitation, a great variety of planting designs might indicate a net competitive effect. Since it is patently impossible to investigate *all* planting designs, one can never be certain that the particular design which would have demonstrated facilitation had really been investigated.

Under certain circumstances, the use of 'target' experiments may obviate this problem. So-called target experiments (Goldberg & Werner, 1983; Vandermeer, 1986; Goodall, 1960; Mead, 1979) take on a variety of forms, but have the common feature that a single individual, the target, is surrounded by a number of other individuals. The 'other' individuals may be randomly positioned within what is thought to be an area of influence, or they may be placed in a systematic arrangement. A convenient form is to place an identical number of individuals at a fixed distance from the target, varying the distance from very close to the target to very far from the target. The target will thus have its yield reduced, presumably as a function of the distance to the competitors.

To eliminate the known effect of biomass of competitors, one can compute a 'competition coefficient', which represents the yield loss of the target per unit biomass of the surrounding competitors. If k is the unencumbered yield (or biomass) of the target individual, $y_t(\Delta)$ is the yield (or biomass) of the target when in competition with surrounding plants at distance Δ, and $y_c(\Delta)$ is the yield (or biomass) of the surrounding competitors at distance Δ, we have

$$\alpha(\Delta) = [k - y_t(\Delta)]/y_c(\Delta).$$

This competition coefficient can then be plotted against the interplant distance (distance from target to surrounding plant $= \Delta$). Provided that enough distances are included in the experiment, this methodology should detect the existence of net facilitation, if it is actually occurring. An example for common bean affecting tomato is presented in Figure 6.5.

Naturally, if the facilitative mechanism is only possible with particular planting designs (e.g. aphids that respond to distinct 'edges' of row-cropped vegetation), such target experiments will not be justified. But I suspect many cases that might be difficult to discover with ordinary plot experiments will be easily demonstrated with a series of target experiments.

7

Special problems in intercrops involving perennials

Introduction

While the focus of this book is on annual crop systems, some of the more common and spectacular examples of intercropping involve perennial crops. Rappaport's classic study of the Tsembaga in New Guinea shows a complex system of succession and intercropping which includes many perennials (Rappaport, 1967). The so-called village forest-gardens in West Java exhibit similar diversity (Michon *et al.*, 1983), as do the compound farms of Nigeria (Okigbo & Lal, 1978). To this day the Maya of Mexico maintain kitchen gardens sometimes growing over 30 species of plants, including fruit trees, bananas, and other perennials, (Alcorn, 1984). In South-East Asia the use of perennials in slash and burn agriculture is well-known (Spencer, 1966).

But it is not only in traditional peasant production that perennials are integral components. Commercial plantations involving intercrops are legion. Coffee and cacao are almost always grown with shade trees (usually a legume), implicitly an intercropping situation, although the production of the shade trees is of no consequence to the producer (Aranguren *et al.*, 1982*a*, *b*). But many examples exist of the joint production of two perennials: cacao and coconut (Aggaoili, 1961; Garot & Subadi, 1958; Jose, 1968; Leach, 1971; Traeholt, 1962), coffee and rubber (Townsend *et al.*, 1964), a variety of examples with African oil palm (Sparnaaij, 1957; Webster, 1969; Blencowe, 1969; Soekarno, 1961; Wood, 1966), and at least seven different perennials with coconuts (Nair, 1983). A common situation in commercial production is the growth of annual row crops in young plantations. For example, in Sri Lanka it is common to cultivate basic grains in young rubber plantations (Senenayake, 1968), and in Malaysia, corn, sorghum, peanut and soybeans are planted in young plantations of African oil palm (Cheng, 1970). Similar biological phenomena are involved with the taungya systems of reforestation (Watson, 1983; Budowski, 1981). Finally, a perennial intercrop that has been observed by all visitors to the tropics is the joint production of a perennial crop and forage (e.g. coconut plantations with their common comensals, *Bos taurus* (Plucknett, 1979)) a system that arguably is often not very well planned.

On the one hand, the ecological generalizations already elaborated should apply regardless of the perennial nature of the species involved. Thus, all the material in this book is in principle just as applicable to intercrops involving perennials as it is to other types of intercrops. On the other hand, there are situations imposed on the system through the perennial crop that are sufficiently unique and homogeneous to justify additional special treatment. For example, as noted below, when annual crops are intercropped in a young plantation the producer is not concerned with the joint maximization of the two commodities (maximizing LER) but, rather, with the maximization of the annual crop production without significantly reducing the growth rate of the plantation species.

Some of these special situations can be visualized in the qualitative chronological sequence of the growth of a coconut plantation (Nair *et al.*, 1979). Figure 7.1 shows such a sequence. In Figure 7.1(*a*) the plantation is young and its sparseness provides ample room for intercropping. At an intermediate growth phase (Figure 7.1(*b*)) there is less room for intercropping of any sort, first because the canopy of the perennial is large enough to completely cover the ground, and second because the height of the perennial is not yet sufficient to allow for the generation of adequate understory light environment. Finally, in Figure 7.1(*c*), the trees are sufficiently large that an understory light environment conducive to adding an intercrop is present. Thus, we are interested in both young and old plantations, but the particular problems associated with each are likely to be quite distinct. The present chapter is organized according to this logic.

The materials will be grouped in two categories, corresponding to the two categories most commonly encountered in nature. The first is the case of cultivation of annual crops in young plantations (as in Figure 7.1(*a*)), particularly with respect to the joint problems of maintaining good growth rates of the perennial while trying to maximize production of the annual crop. The second is production of an understory crop in a mature plantation (as in Figure 7.1(*c*)). I hasten to add that this chapter in no way is meant to summarize the enormous literature on agroforestry (e.g. Huxley, 1983), much of which could probably be reviewed in the context of previous chapters.

While the following is, as anticipated above, quite eclectic, there are nevertheless some general principles upon which most of the developments rest. First, with regard to the production of annual crops in a young plantation, the general idea of the competitive production principle is directly applicable, with the important change that the yield of the plantation crop is not really of interest, but, rather, the growth rate of individual trees. Thus, rather than comparing the yields with intercrops to those resulting from monocultures, the interest is in balancing a good yield of the annual crop with an acceptable growth rate of the plantation species.

Fig. 7.1. Illustration of the underplanting potential of a palm plantation at different stages of development: (*a*) very young plantation in which ample space between trees permits understory cultivation; (*b*) intermediate stage of plantation in which the canopies are well-developed but the trees not very high, creating less than ideal conditions for underplanting; (*c*) a mature plantation in which sufficient light enters the understory to make conditions again suitable for underplanting (redrawn from Nair, 1983).

Second, the treatment of an understory crop in a plantation can be formulated as an application of the competitive production principle, in the specific context of effect–response, where the environment is the light environment to which the understory species will be subjected and the effect is the qualitative and quantitative pattern created in the understory by the perennial. The basic idea is that the crown of a tree intercepts some, but not all, of the available incident radiation, and thus modifies the environment of the understory species. Since not all light is intercepted, some is available to any plant which happens to be in the right position below. It is a very simple idea, but one that can be exploited to some extent by the theoretician. As a remarkably ingenious exploitation of this basic idea, consider the following intuitive description of the geometry of an individual plant by Horn (1971):

> Since the curve of net photosynthesis saturates at 20% of full sunlight, any leaves that are lit more intensely must operate at full capacity. Thus we can imagine a layer of leaves with one unit of leaf area for every 2 units of ground area, operating at the peak rate in full sunlight. . . . [at some distance] below this layer, the light intensity is reduced to 50% of full sunlight, and we can imagine a second layer of leaves, one leaf for every 2 units of ground area, again operating at peak photosynthetic rate. Finally [at yet a further distance] below . . . the light intensity is 25% of sunlight, and since this is just above the saturation point for photosynthesis, we can cover the available ground area with a third layer of leaves, all operating at peak photosynthetic rate. Thus we design a multilayered strategy that exposes ($1/2 + 1/2 + 1 = 2$) twice as much leaf area as there is ground area, with all leaves operating at peak photosynthetic rate.

Just as this elegant analysis explains why plants are three-dimensional, the same ideas can be manipulated to suggest how multidimensional intercrops should be structured. This idea, in much disguised form, is the rationale for examining the light environment in the understory of a mature plantation. The approach is based on two simple biological principles, (1) photosynthesis is systematically related to radiation intensity, and (2) a particular plantation planting creates a predictable pattern of understory radiation.

Young perennial plantation with an annual intercrop

A common situation encountered wherever plantation crops are important is the underutilization of land during the time the plantation is in its initial growth stages. While it is frequently the case that such situations are exploited with annual crops (e.g. Senenayake, 1968; Cheng, 1970), the case is rarely treated either experimentally or theoretically. Basic biological facts suggest that utilizing unutilized space in a young plantation would be a beneficial use of land, yet the principal agronomic goal of a young plantation is to obtain the maximum growth rate of the trees, a goal that might be hindered by the competitive presence of an annual crop in the system. The problem is thus to determine the balance between production of the annual crop and reduction in growth rate of the tree crop. More specifically, the producer must know how much time will be lost in getting the plantation to its productive size. We thus seek a formulation that relates the time necessary for the plantation to reach production to the yield of the annual crop. Naturally, if the annual crop has a negligible effect on the growth of the perennial, the approach developed here is superfluous and the problem is simply one of determining optimal design of the annual. There are, however, at least a few examples in the literature in which undercrops have been shown to have a significant negative effect on the growth of young plantation species (e.g. Redhead *et al.*, 1983; Agamuthu & Broughton, 1985).

Suppose that y is the biomass (per hectare) of the perennial crop (which we call 'tree' from this point on) and r is its rate of growth. Supposing a simple exponential increase, which is a reasonable assumption for initial stages of growth, we can write

$$y_t = e^{rt} \tag{7.1}$$

Let us further make the reasonable assumption that r depends on the density of the annual crop. Assuming initially that r is a linear function of the density, we denote the density of the annual crop as D and write

$$r = r_0 - Dr_0/D_0, \tag{7.2}$$

where r_0 is the value of r when $D=0$ and D_0 is the value of D when $r=0$.

Supposing that x is the yield of the annual crop (what we call simply the 'crop' henceforth in this section), recall the basic relationship between density and yield (Vandermeer, 1983; Bleasdale & Nelder, 1960; also see Chapter 10) as

$$x = aD/(1+bD^c), \tag{7.3}$$

where a is the unencumbered yield (yield of an isolated crop plant), b reflects the overall intensity of competition, and c is a measure of how rapidly competitive effects decay as the distance between competing plants increases (see Vandermeer, 1983, for a complete discussion of the meaning of these parameters). It is reasonable to assume, in most cases of interest here, that c is approximately equal to unity, an assumption we make for the rest of this section, and one that is relaxed in Chapter 10.

Suppose that y_c is the 'critical biomass' of the tree, that is, the biomass which the tree must attain to begin producing. Therefore, if T is the time it takes the tree to reach this point, we can write (from equation (7.1))

$$T = r^{-1} \ln y_c.$$

Substituting for r in equation (7.2) and simplifying, we obtain

$$D = (r_0 T - \ln y_c)/(Tr_0/D_0),$$

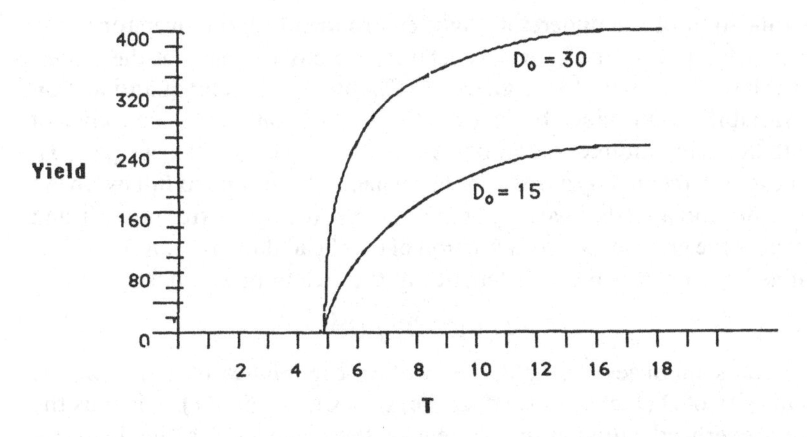

Fig. 7.2. Critical function for planning the production of understory annuals in a young plantation. Yield of annual crop is represented as a function of the time necessary for the perennial to reach maturity, for two values of the parameter D_0 (see text) (redrawn from Vandermeer, 1983).

which we substitute into equation (7.3) to obtain the critical equation

$$x = (ar_0T - a\ln y_c)/[(r_0/D_0 - br_0)T - b\ln y_c], \qquad (7.4)$$

which relates the yield of the crop to the time necessary for the plantation to reach productive size. Equation (7.4) is shown for various values of the parameter D_0 in Figure 7.2. Note the general form of the relation. As we would expect qualitatively, the yield of the crop increases rapidly for small values of T and slowly for larger values.

If the parameters of equation (7.4) can be estimated in actual cases, a task that usually should not be difficult (Vandermeer, 1983), equation (7.4) could be quite useful in planning row intercrops in young plantations. With economic information added, particular crop yields can be converted to their monetary value and the associated time lost in production of the perennial (T) can also be converted into its monetary equivalent. A simple economic comparison then can establish whether it makes sense to plant the row intercrop in the first place, and, if so, at what density. A major problem with this formulation is that the parameters of the critical function are likely to change with different stages of the perennial species, necessitating the estimation of parameters specifically for each stage of the perennial.

The light environment in a mature plantation

Some qualitative features

The production of an understory crop, whether it be a continuous ground cover crop or a different, structurally distinct, species of crop, is conditioned by the modification of the understory light environment by the overstory crop (e.g. Newman, 1986). The overstory 'effects' the environment of the understory, the latter 'responds' to the affect (see Chapter 3). A useful quantification of this variable was provided by Jackson (1983). Jackson's model conceives of the total direct and indirect radiation as divisible into two parts, one part, T_f, that is never intercepted by a tree and thus reaches the ground as full overhead illumination, and a second part T_0, that is intercepted by the tree crowns, and thus reaches the ground as only a fraction of overhead illumination. T_c will be determined by the crown characteristics of the tree in question, so that

$$T = T_f + (1 - T_f) f(D),$$

where $f(D)$ symbolizes the light interception capabilities of a tree with a crown density of D (Jackson also discusses the exact form of f). T is thus the fraction of overhead illumination that reaches the ground and is thus available for an understory crop.

As a sort of average technique, Jackson's approach would seem to be most useful in an already extant plantation, for making decisions about the feasibility of underplanting particular species. But for other applications, such as designing a plantation from scratch or planning the details of exactly where in the plantation an understory should be planted, a somewhat different tact will be taken here.

First consider the shadow cast by a single layered object at different times of the day, as presented in Figure 7.3(*a*). The shadow image is the same size as the object whether at 10 a.m., or noon, or any time. The shadow image of a solid object casts a different size shadow depending on the time of day, as illustrated in Figure 7.3(*b*). Thus, depending on the shape of the overstory tree, different amounts of shade will be cast on the ground. In Figure 7.3(*a*), the ground position labeled *p* was in full sun at 10 a.m., in shade at 11 a.m. and again in sun at noon, thus spending an hour or less in shade. The same point in Figure 7.3(*b*) was in shade at 10 a.m., remained in shade at 11 a.m. and only by noon was it exposed to full sun. It is thus easy to see how the shape of the overstory tree, and not only the density of its leaves, partially determines the overall amount of direct radiation passing through to the ground (or some other level).

In Figure 7.4(*a*) tree shadows are illustrated for three different times of day, based on measured shadows of *Cressentia alata* in northern Nicaragua (Vandermeer & Meyrat, 1987). If we construct a theoretical plantation of these trees on a grid such that the trees are 6.5 m apart, the shadows at ground level would be expected to be as in Figure 7.4(*b*). The light gaps that one commonly sees in plantations are illustrated by the stippled areas. While it is quite obvious where the light gaps are located and how they are formed, it is also true that these gaps are constantly moving as the sun traces its trajectory

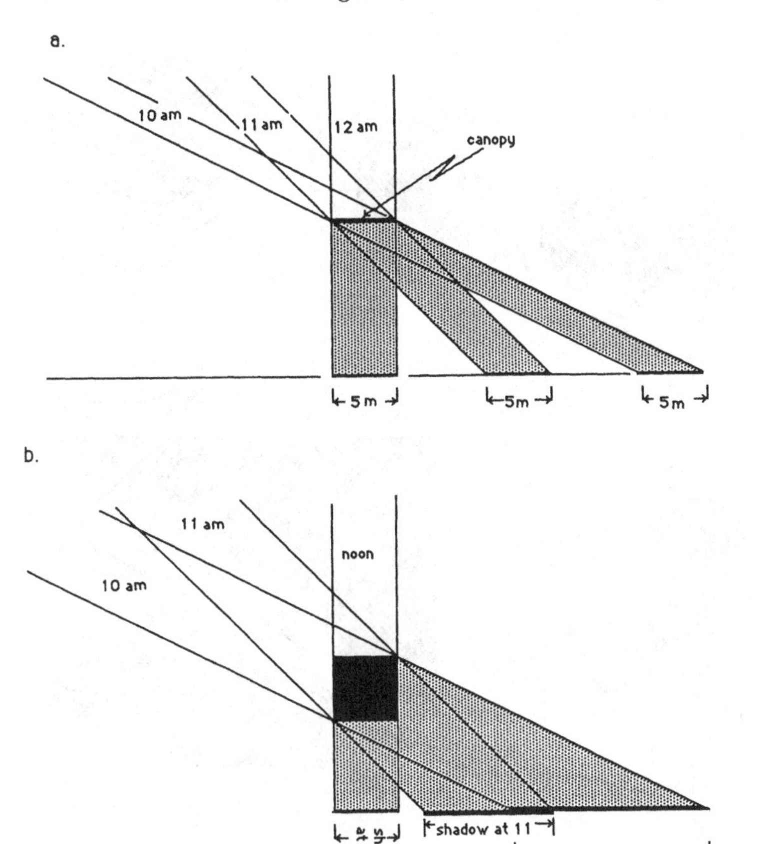

Fig. 7.3. Shade-casting properties of a solid object (*a*) a two-dimensional object; (*b*) a three-dimensional object.

through the sky, and that they are changing size from small early in the day to their maximum at noon to small later in the day (Figure 7.4(*c*)). Most important we note that with the exception of a small area directly under the tree, the rest of the understory area is subjected to direct radiation at least part of the day, a nontrivial observation. For example, the diagram in Figure 7.5(*a*) shows the position of a single tree's shadow during the sun's main daily trajectory. The shadow cast by the tree changes from an ellipse to a circle (in this idealized case) and back to an ellipse. But, more importantly one can see how various areas of the understory can be expected to experience different

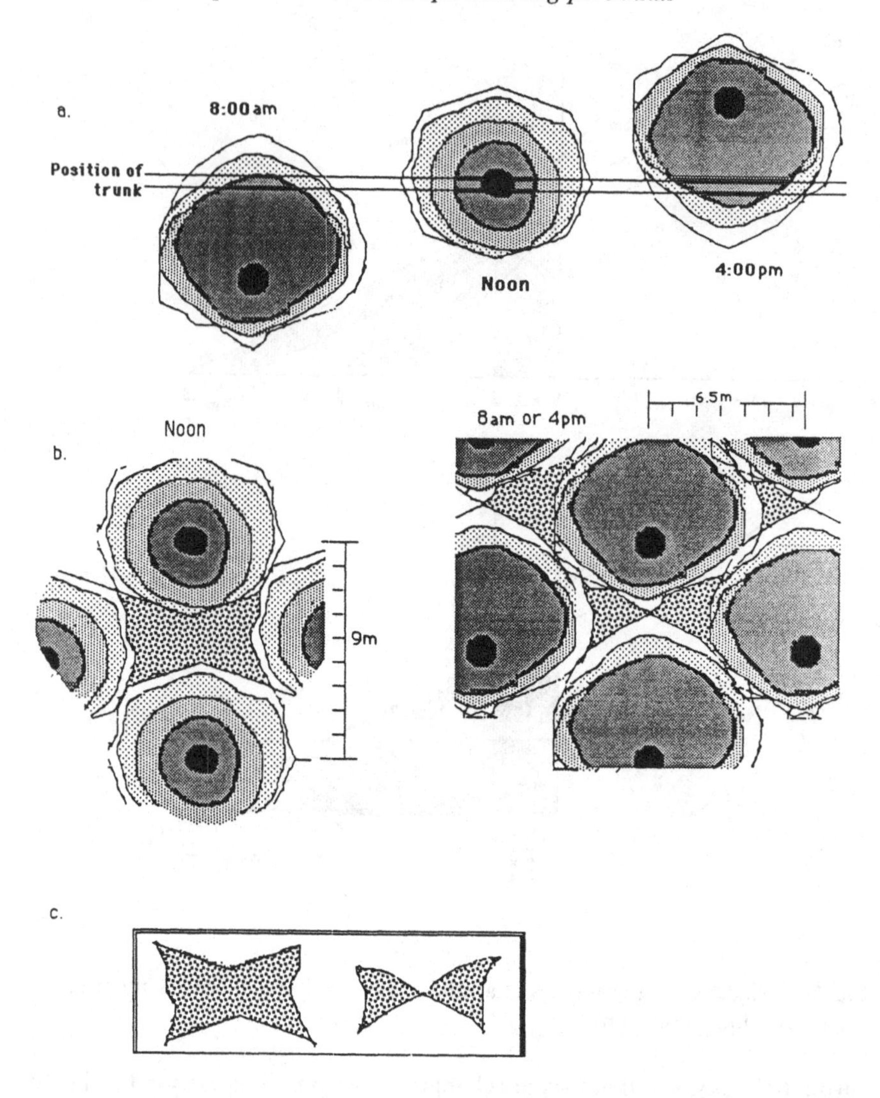

Fig. 7.4. Shadows cast by a *Cressentia alata* tree in northern Nicaragua. Density of shading in diagram is relative to degree of shade in the shadow. (*a*) Positions of shadows at three times of day. (*b*) Shadow pattern in a hypothetical plantation at noon and at 8 a.m. (or 4 p.m.). Trees are arranged in a square lattice with 6.5 m separating each tree. Stippled area represents a light gap. (*c*) Characteristic light gaps at noon (left) and in morning or afternoon (right).

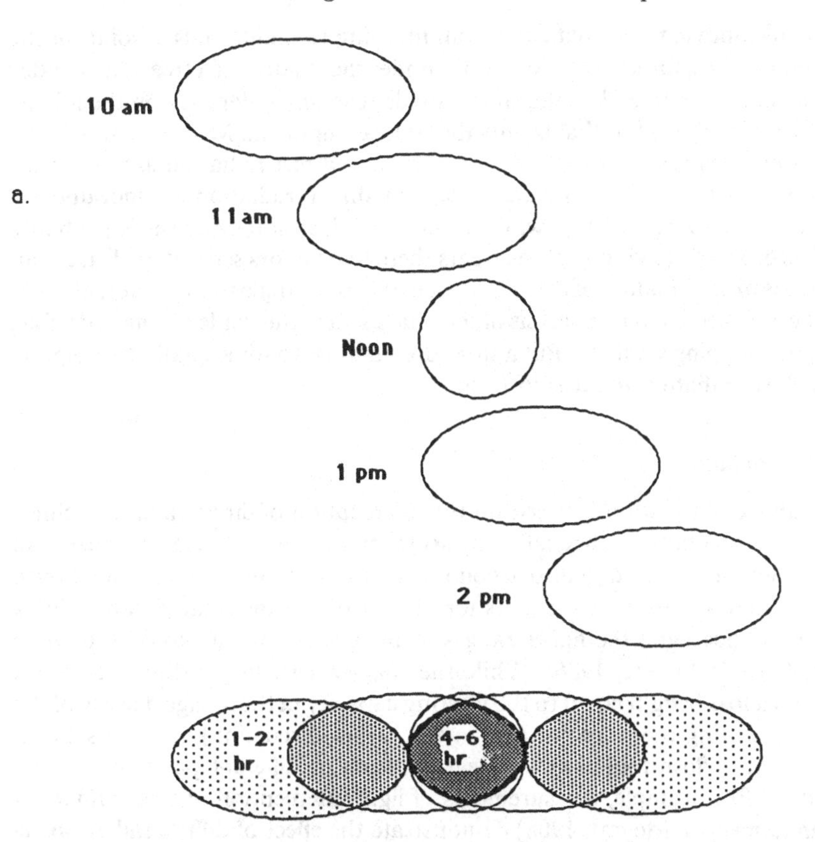

Fig. 7.5. (*a*) Theoretical positions of idealized shadows. (*b*) Construction of the effective shadow, illustrating relative amounts of time a point is expected to remain in the shadow.

amounts of sunlight during the course of the day, as shown in Figure 7.5(*b*). Relatively far from the tree the ground experiences shade for only one hour during the day, while directly under the tree, shade is experienced for three hours. This creates an area on the ground that is more-or-less ellipse-shaped, where the relative concentrations of sunlight (as a consequence of the various shadow patterns created during the day) are shown with different degrees of shading in Figure 7.5(*b*). We also note that the actual amount of radiation falling on this area, as well as the entire region of effective shade, depends on the density of leaves in the canopy of the overstory tree.

The general pattern to be expected is a non symmetrical distribution of light interception, more skewed as the time of measurement deviates from noon. But from the point of view of a plant in the understory, the shadows are dynamic, moving systematically from morning to night. Thus a point on the ground is (1) in direct radiation, or (2) under the shadow of a tree. If it is under the shadow of a tree, the intensity of shade reaching it depends on the foliage density of the tree. But that is only the static component. More important, the point experiences a certain amount of time in direct radiation and a certain amount of time in the shade. A leaf in direct radiation is undoubtedly photosynthesizing at 100% while in the shade its photosynthesis is probably reduced. The following analysis treats these two factors separately. First is an analysis of the 'quality' of the shade produced by a single tree at different times of the day. Second is an analysis of the time spent in the shade of single, double, and overlapping shadows. But a short digression into some qualitative aspects of diffuse radiation input is in order.

Diffuse radiation

The above description is based on the interception of direct radiation. Since much of the photosynthetically important radiation is diffuse, we must take that fraction of radiation into account also. For example, in temperate South Africa, diffuse radiation accounts for about 30% of the total (Chang, 1968), while in Nicaragua the figure ranges from 16% in January to 45% in June (Lopéz de la Fuente, 1986). While the shadow cast in the direct radiation implies a loss of light equal to the filtering capacity of the foliage density of the canopy casting the shadow ($f(D)$ in the earlier discussion), it is possible to detect the diffuse radiation by measurements in different parts of the cast shadow. For example, measurements of light interception of *Cresentia alata* (Vandermeer & Meyrat, 1988) demonstrate the effect of diffuse radiation, as shown in Figure 7.6. In Figure 7.6(*a*) the relative light penetration is shown at noon (along with a diagrammatic representation of the overhead canopies). As can be seen there are two shadows clearly present, one beneath each tree. The light gap in between the two trees is pronounced. In Figure 7.6(*b*) the same two trees are shown at 3.00 p.m. Again the two shadows are there, but the pronounced light gap is at best just a suggestion. Why? The light gap at this point (on a transect in the line of solar movement), while smaller in absolute size, should be just as intense as at noon. The most obvious interpretation is the loss of the diffuse radiation input. Adding credence to this interpretation is the fact that the 'tail' of the shadow of the tree on the right is significantly brighter than it was earlier, again not to be expected simply from the model of intercepting direct radiation.

The above suggests that a shadow, at a particular point in time, has a pattern of variation in fraction of incident radiation penetrating to the ground, and that pattern is determined not only by the thickness of the tree crown which intercepts the direct radiation, but also by the amount of diffuse

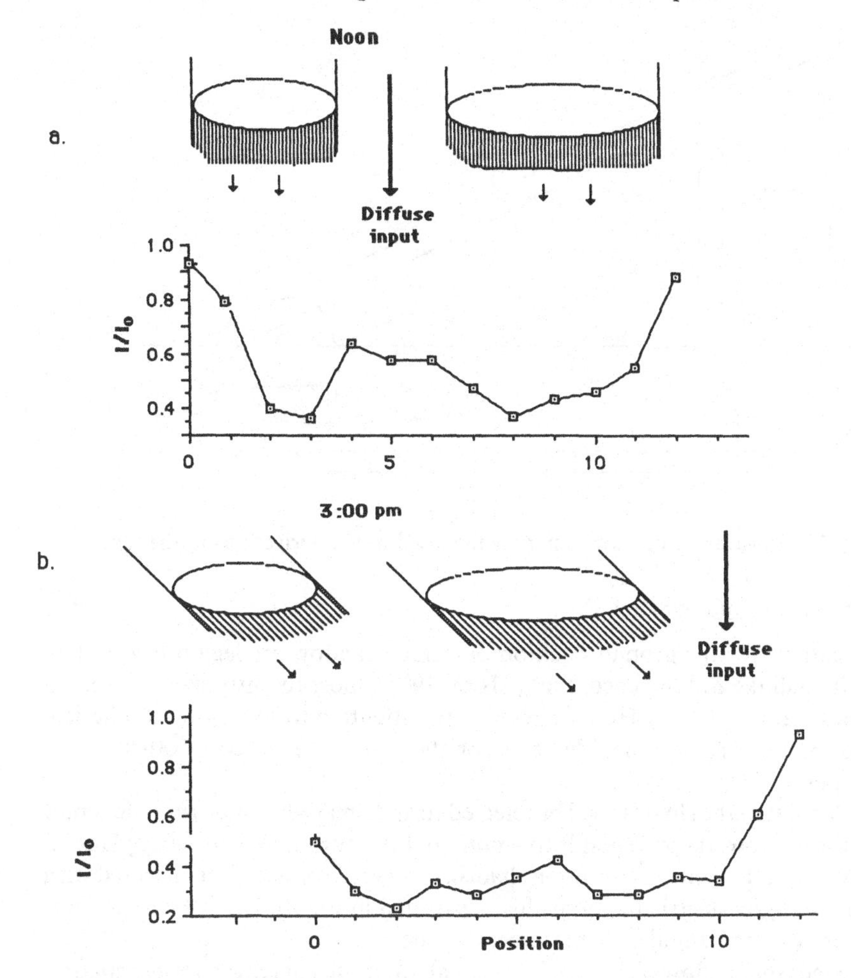

Fig. 7.6. Empirical measurements of light penetration in an isolated pair of *Cresentia alata* trees in northern Nicaragua. (data from Vandermeer & Meyrat, 1988).

radiation penetrating to the shadow (this also explains why an older coconut plantation receives a greater amount of light at ground level than a middle-aged plantation, as represented in Figure 7.1). In the end we must thus consider three variables that dictate the effective intensity experienced by an understory plant: the interception pattern of the tree crown, the time-course of the shadows, and the diffuse radiation input. The following two sections refer to the first two of these variables. The analysis of the diffuse input awaits development.

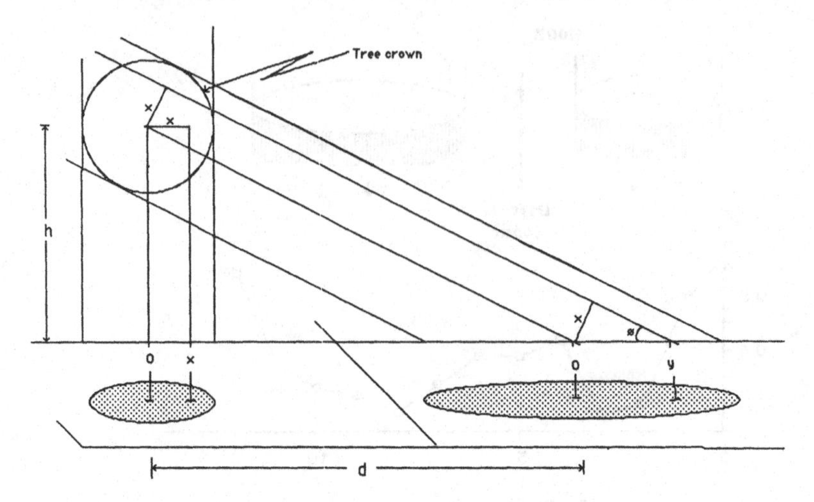

Fig. 7.7. Illustration of various parameters used in the derivations in the text.

The shadow of a canopy

Equations for the ground position of a tree's shadow are legion (e.g. Usher, 1970; Jahnke & Lawrence, 1965; Horn, 1971), most recently summarized by Quesada *et al.* (1987). Here we restrict our attention to a tree with a spherical crown, referring to Quesada *et al.* for the reader interested in other crown shapes.

A point in the shadow will be referred to as displaced y units from the center of the shadow. It corresponds to a point in the noon-time shadow displaced x units from the center of the noon shadow. The value of y can be computed with simple trigonometric relationships, illustrated in Figure 7.7, to be $y = x/\sin\phi$, where ϕ is the angular displacement of the sun.

Since the shadow is an ellipse (except at noon when it is a circle) its equation is simply that of an ellipse whose main axis is the radius of the crown divided by the sine of the angle of displacement. If z_1 and z_2 represent displacement from the center of the shadow along the two principle axes, we have

$$z_1^2/r^2 + z_2^2\sin\phi^2/r^2 = 1$$

as the equation for the shadow.

The shadow itself will be displaced $h/\tan\phi$ units from the center of the tree, where h is the height of the tree, as shown in Figure 7.7.

The shadow with a filtering canopy

As with any vegetation, the canopy of the tree will tend to act as a filter, reducing the amount of light penetrating to a given level in proportion to the

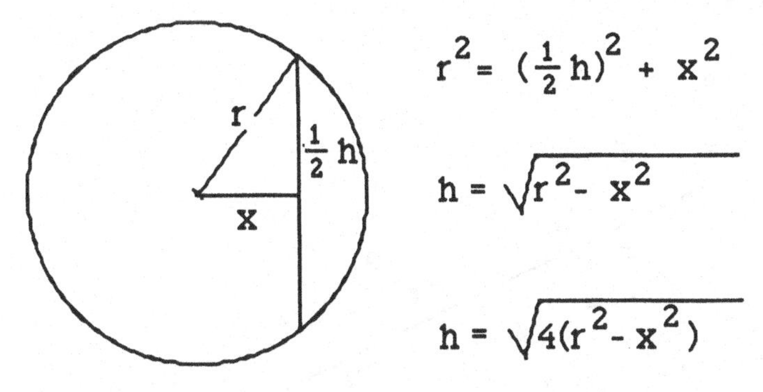

Fig. 7.8. Computation of the length of the chord at position x, displaced from the center of the crown of the tree.

vegetation density at that level. For theoretical reasons, the light intercepted by vegetation with a leaf area index of L is given as

$$I = I_0 e^{-kL},$$

where I is the radiation, I_0 is the open sky radiation and k is a constant. This relationship is known as Beer's law and has been confirmed, at least approximately, in several cases (Saeki, 1960; Monsi & Saeki, 1953). An alternate expression, more complicated, was proposed by Monteith (1965) and yet another by Horn (1971). But for the purposes of this chapter, the work of Stanhill (1962) is more appropriate, in which the fraction of radiation beneath a crop was a simple linear function of the crop height. Thus

$$I/I_0 = 1 - bh, \qquad (7.5)$$

where b is a constant and h is the thickness of the vegetation. Depending on the relationship between vegetation thickness and leaf area index, Beer's law could easily be approximated by this linear form.

For purposes of the present development, the thickness of the vegetation is the same as the thickness of the crown of the tree, and we are interested in these values for particular values of x (displacement from center of noon-time shadow) and eventually y (displacement from center of some other shadow). At point x in the noon-time shadow, the chord displaced x units from the center of the tree crown, has a length

$$h(x) = [4(r^2 - x^2)]^{\frac{1}{2}}, \qquad (7.6)$$

as can be seen in Figure 7.8. Substituting equation (7.6) into (7.5) we obtain

$$I/I_0 = 1 - b[4(r^2 - x^2)]^{\frac{1}{2}},$$

which, after rearrangement and substituting for $x = y\sin\phi$, we obtain

$$(1 - I/I_0)^2 = (2rb)^2 - (2b)^2(y\sin\phi)^2, \qquad (7.7)$$

which is linear in $(1 - I/I_0)^2$ and $(y\sin\phi)^2$. Thus, from simple ground

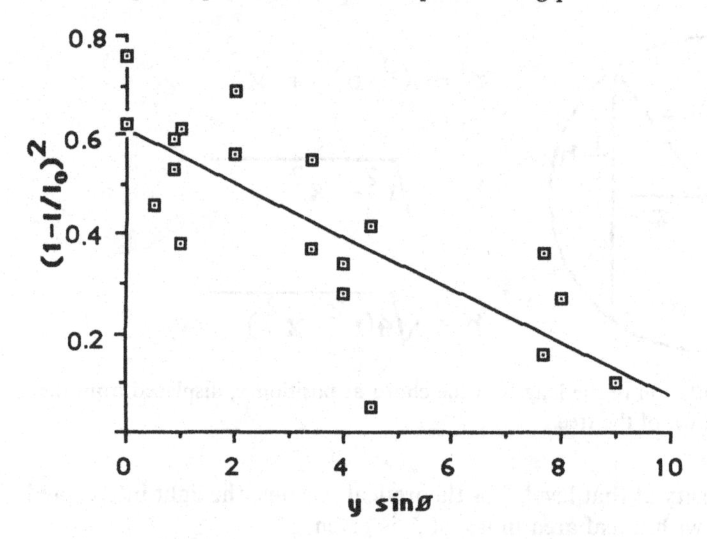

Fig. 7.9. Empirical measurements of equation (7.7); data from an isolated *Cresentia alata* tree in northern Nicaragua (Vandermeer & Meyrat, 1987).

measurements both parameters, r and b, can be measured. Data from *Cresentia alata* (Vandermeer & Meyrat, 1988) are shown in Figure 7.9, corresponding to a linear fit. Based on this linear approach, the prediction of relative intensities of shadows at three different times of day are shown in Figure 7.10.

Finally, note that all of this derivation is based on points arranged on a line corresponding to the reflected path of the sun. Lines displaced a certain distance from the sun, but parallel to its main axis, require a slightly modified formula whose derivation is a simple trigonometric exercise.

Time in the shadow

As a problem distinct from the previous, now consider the problem of the time that a point is in the shade. With reference to Figure 7.11, we see graphically a point intercepted by the shade of a tree, and later left by the same shadow. The time that the point remained in this shadow can be expressed as an angle, which we call β. Thus, time is expressed as degrees of arc (the conversion to real time is accomplished by multiplying the angular time by 4, since there are 4 minutes in each degree).

Figure 7.11(b) shows the variables necessary to derive an equation for the angle β. By inspection we see that $\sin(\beta/2)=r/b$. The value of b can be derived

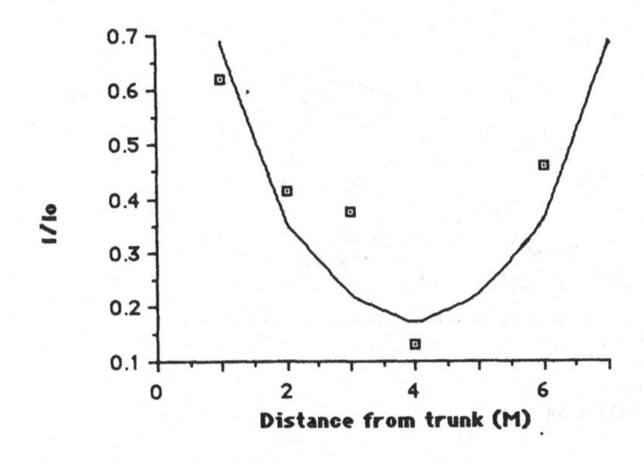

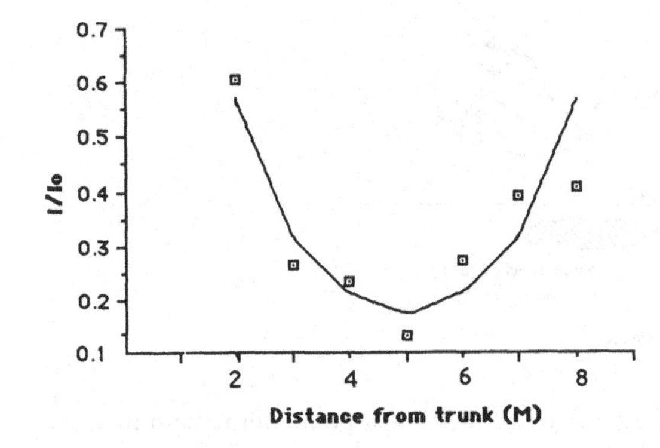

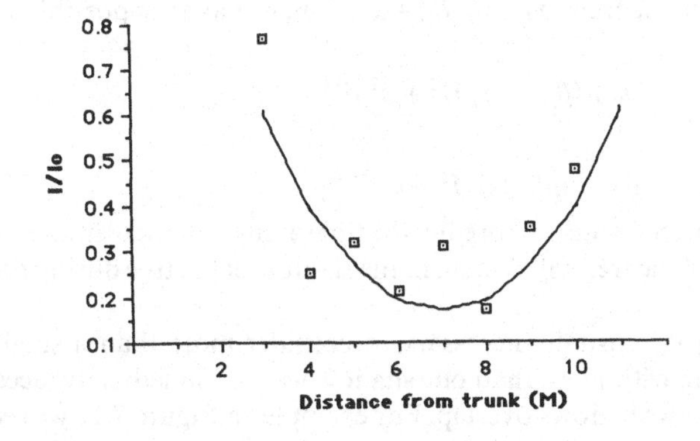

Fig. 7.10. Predictions of relative light intensity at ground level below an isolated *Cresentia alata* tree in northern Nicaragua (Vandermeer & Meyrat, 1988).

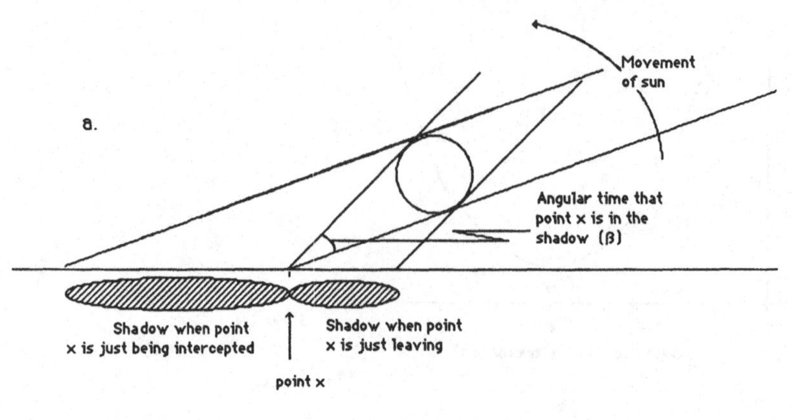

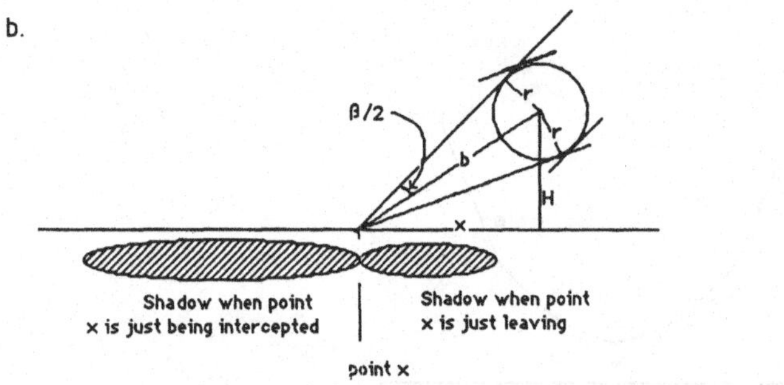

Fig. 7.11. Diagrammatic representation of the time a point will remain in a moving shadow (see text).

from the Pythagorian theorem as $b^2 = H^2 + x^2$, which makes it possible to write

$$\sin(\beta/2) = r[H^2 + x^2]^{-\frac{1}{2}},$$

or

$$\beta = 2\sin^{-1}[r(H^2 + x^2)^{-\frac{1}{2}}]. \qquad (7.8)$$

Equation (7.8) is an exact formula to predict the time a single point displaced x units from the trunk of the tree will remain in the shadow of the tree during the day.

In a real situation it is usually necessary to consider more than a single shadow. When dealing with more than one shadow we are immediately faced with the question of how shadows overlap. For example, in Figure 7.12 we see the situation with two trees. The time in the total shade is simply the time in the shade of the first plus the time in the shade of the second, or

$$\beta = 2\sin^{-1}[r(H^2 + x^2)^{-\frac{1}{2}} + 2\sin^{-1}[r(H^2 + x'^2)]^{-\frac{1}{2}}, \qquad (7.9)$$

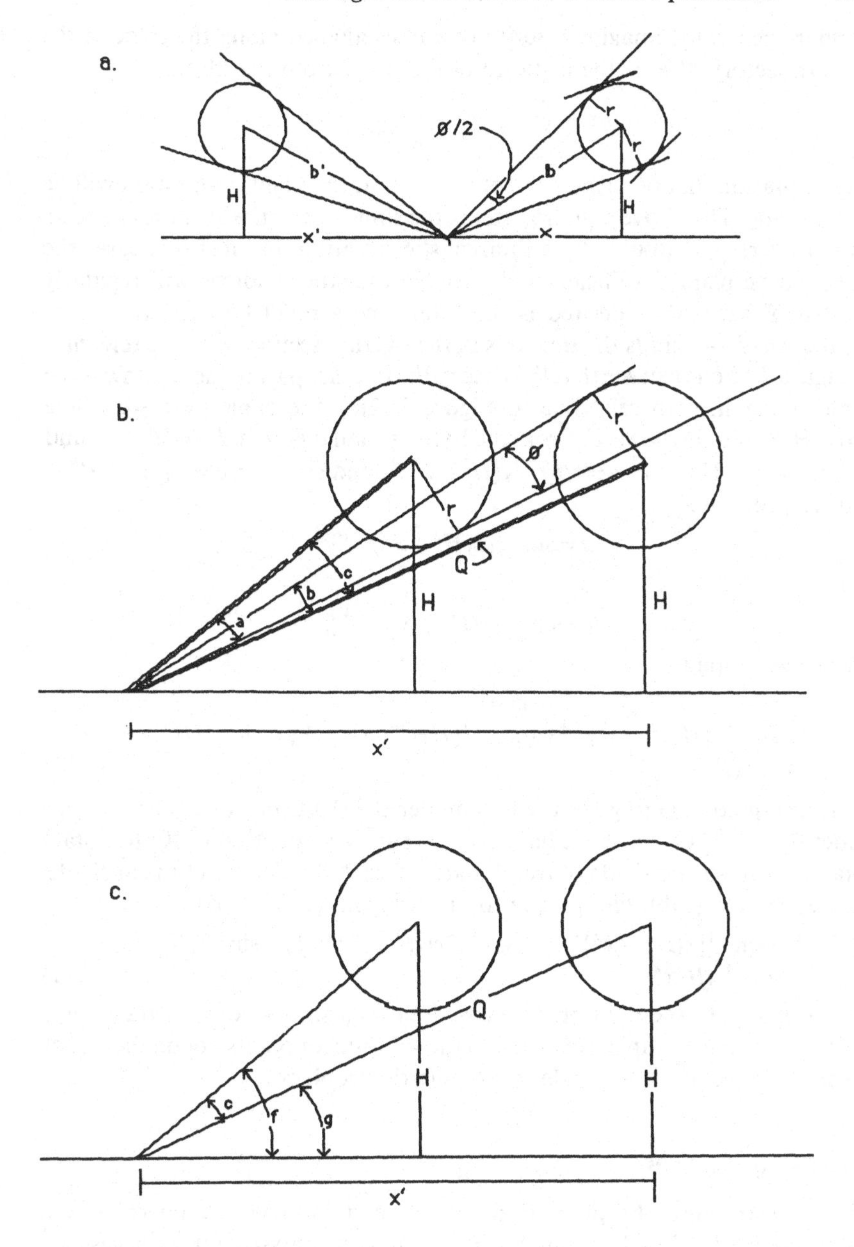

Fig. 7.12. Diagrammatic representation of the time a point will remain under shade when two shadows are involved.

or, more generally, imagine a series of n trees aligned along the trace of the sun's trajectory. If x_i is the distance of the point from the ith tree,

$$\beta=2\sum_{i=0}^{n}\sin^{-1}[r(H^2+x_i^2)^{-\frac{1}{2}}]$$

is the equation that describes the total time that the point in question will be under shade. The above equation treats the simple case in which the shadows do not overlap, undoubtedly an unrealistic situation. In most real cases the trees will be planted sufficiently densely so that the shadows will regularly overlap. What is thus desired is the total time a point falls within one or another shadow, minus the time it is in the overlap section. Making reference to Figure 7.12 one can see that the time in the overlap part of the shadow is the angle ϕ. Again with reference to Figure 7.12(c), the angle $c=f-g$, where $\sin f=H/x$, and $\sin g=H/x'$. From the Pythagorian theorem, $P^2=H^2+x^2$ and $q^2=H^2+x'^2$. Thus $\sin a=r/p=ry(H^2+x^2)^{\frac{1}{2}}$, and $\sin b=r/q=r/[H^2+x'^2]^{\frac{1}{2}}$ and we can write

$$a=\sin^{-1}[r(H^2+x^2)^{-\frac{1}{2}}]$$

and

$$b=\sin^{-1}[r(H^2+x'^2)^{-\frac{1}{2}}],$$

whence we obtain

$$\begin{aligned}\phi&=a+b-c\\&=\sin^{-1}[r(H^2+x^2)^{-\frac{1}{2}}]+\sin^{-1}[r(H^2+x'^2)^{-\frac{1}{2}}]-\sin^{-1}(H/x)+\\&\sin^{-1}(H/x')\end{aligned}\qquad(7.10)$$

Since equation (7.9) gives the time under the shade of tree 1 plus the time under the shade of tree 2, we have only to subtract equation (7.10) to obtain total time under the shade of tree 1 and/or tree 2, that is, total time in shade, thus we obtain (subtracting equation (7.10) from (7.9), $\beta-\phi$),

$$\begin{aligned}T=&\sin^{-1}[r(H^2+x^2)^{-\frac{1}{2}}]+\sin^{-1}[r(H^2+x'^2)^{-\frac{1}{2}}]+\sin^{-\frac{1}{2}}(H/x)\\&-\sin^{-1}(H/x')\end{aligned}\qquad(7.11)$$

The extension to the multiple tree case follows a similar logic as the earlier derivation of the multiple tree case and the solution for points not on the line of the sun's trajectory is a simple trigonometric problem.

Potential applications

The basic structure of a plantation imposes certain inevitable microhabitat features at the level of the ground. It is through an analysis of these necessary features that we can gain an idea of how the production of the ground cover crop works, and thus how we can improve it. The previous section derived exact expressions for the shade-casting properties of a plantation of trees. This analysis could easily form the basis of a computer algorithm (for example) which would predict understory light quality for arbitrarily designed plantations. But the complexity of the light environment is such that one is tempted to attempt a more agglomerated approach to the problem. That is exactly what has been done in the past (e.g. Jackson, 1983).

The agglomerated approach may in the end be the easiest and most effective technique, especially in dense plantations. But I rather suspect that in many cases plantations are normally grown sufficiently sparse that the more disarticulated approach presented above could prove useful. Once the effect of the overstory species is described, each point in the understory is completely characterized with respect to the daily light environment (aside from the sticky question of diffuse radiation). Coupling this information with the 'response' information, that is, translating the characteristics of the 'effected' light environment into the photosynthetic 'response' of the understory species, one could develop a relatively complete model for the design of undercroping systems.

For example, several authors (e.g. Monteith, 1965; Horn, 1971) have noted that the process of photosynthesis can be approximately modeled with the classical Michaelis–Menton equation of enzyme kinetics. Following this procedure we have

$$Z=(Z_{max}p)/(p+k), \tag{7.12}$$

where Z is production (either gross or net, depending on intended use of the model) of the ground cover crop, p is relative radiation intensity (I/I_0 of equation (7.7)), Z_{max} is the maximal possible production and k is a constant specific to the species in question, and equal to the value of p for which Z is $\frac{1}{2}$ the value of Z_{max}. Since p is the relative radiation intensity, equations (7.7) and (7.12) can be combined to predict photosynthesis of the understory at point y distance from the tree at a time when the sun is ϕ degrees in the sky. When not in the shade, as computed by an expanded form of equation (7.9), the understory will be at maximum photosynthetic rate, producing Z_{max} for each appropriate time unit. One might thus develop a computer program with the following algorithm (for example):

1. compute time out of shadow (equation (7.10));
2. compute total photosynthesis for time out of shadow (Z_{max});
3. compute position in shadow (equation (7.7));
4. compute photosynthesis for that position (equation (7.12)).

Steps 3 and 4 would be iterated over the whole day. This entire procedure would be repeated for each point on the ground, thus mapping out a potential photosynthetic surface for the understory of arbitrary plantation designs (ignoring difuse input).

But this is only a rough outline. In developing an algorithm for a real situation the equations would have to be elaborated more completely to take into account sun trajectories not on line with the main axes of the trees, different solar inclinations to correct for latitude, and different land slopes. None of these issues creates analytically difficult problems. The question of diffuse radiation, however, does present a serious problem.

Finally, a cautionary note is in order. Thus far we have tacitly assumed that the understory has little or no competitive effect on the plantation species. In an earlier section we examined the case of a young plantation in which the whole point of the analysis was that this assumption is *not* valid. In the case of a mature plantation the assumption is probably usually valid, but even if not, the methods of former chapters (including the simple quantitative ones of Chapter 2 and the more theoretical ones of subsequent chapters) are quite applicable here. Once it is established that the understory crop is a valid possibility, the use of land equivalent ratios, relative value totals, or other methods may be applied directly to finally decide as to the viability of the combination. What is useful about the plantation–understory crop situation is that because of the obvious modifying effect of the plantation species' necessary shading, one can begin to make some *a priori* predictions about intercrop combinations, something which is not so obviously approachable in combinations of annual intercrops, or structurally similar perennials.

8

Weeds and intercrops

Weed control is often cited as one of the benefits of intercropping (Moody, 1980; Shetty & Rao, 1979; Unamma *et al.*, 1986; Robinson & Dunham, 1954; Ibgozurike, 1971; Liebman, 1986). The presumed mode of action is that one crop, through competition with the weed, provides an environment of reduced weed biomass for the other crop. An apparent example is presented in Figure 8.1, in which a luxuriant growth of *Amaranthus* sp. in corn, is dramatically suppressed by a secondary crop of beans. In effect, the beans seem to have been able to replace the *Amaranthus* completely, and either provide a source of valuable yield, or compete less with the corn than the *Amaranthus* does.

Perhaps the best-known example is the use of 'cover crops', between rows of a monoculture. Liebman (1986) reviewed nine studies involving 23 crop–cover-crop combinations. Of the 23 cases, all but three showed a significant weed suppressive effect.

While the literature on cover crops is impressive, and suggests considerable advantage is to be gained in weed control, the literature on combinations of two crops is less extensive and more equivocal. For example, Ayeni *et al.* (1984), working in Nigeria, found that a maize–cowpea intercrop failed to significantly suppress weeds in the early cropping season but had a significant effect in the late cropping season. Unamma & Ene (1983) failed to find weed suppression in a cassava–maize system in Nigeria, whereas Soria *et al.* (1975) found this combination successfully suppressed weeds in Costa Rica. In an extensive literature survey, Liebman (1986) found that weed suppression was stronger in intercrop than in the monocultural components in eight cases, intermediate between monocultural components in eight cases, and weaker than all monocultural components in only two cases (Table 8.1).

The addition of a weed or weed complex to an intercropping situation creates an interesting system from an ecological point of view, a system of three interconnected competitors. Recent ecological theory (Levine, 1976; Vandermeer, 1980*b*; Vandermeer *et al.*, 1985; Levins, 1975) was developed specifically for dealing with this situation. It is effectively an extension of the idea of the competitive production principle, but with the complications involved with a

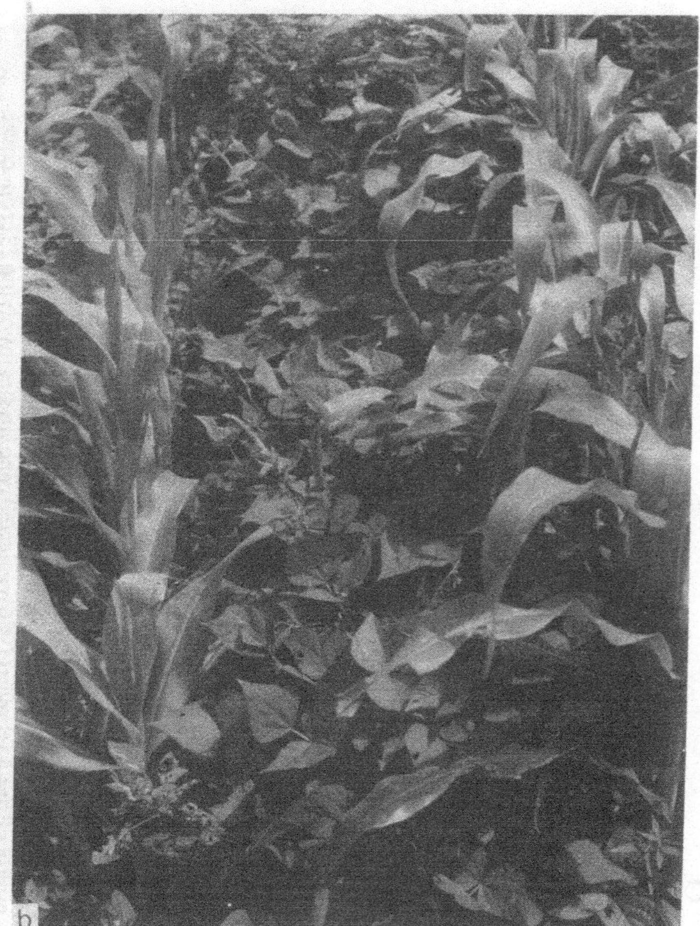

Fig. 8.1. An intercrop of maize and beans in Nicaragua: (*a*) maize monoculture with luxuriant growth of

Table 8.1. *Strength of weed suppression effects by intercrops in which all component crops are considered 'main crops'*

	Weed-suppression effect		
Intercrop combination	Stronger than monocultures of all components	Intermediate between monocultures of components	Weaker than monocultures of all components
Maize–bean	Fleck *et al.* (1984)	Soria *et al.* (1975)	
Maize–cassava	Soria *et al.* (1975)		Soria *et al.* (1975)
Maize–bean–cassava	Soria *et al.* (1975)	Soria *et al.* (1975)	
Maize–mung bean	Bantilan *et al.* (1974)		
Maize–sweet potato		Bantilan *et al.* (1974)	
Maize–peanut		Bantilan *et al.* (1974)	
Maize–sunflower	Fleck *et al.* (1984)		
Maize–cowpea			Ayeni *et al.* (1984)
Bean–cassava	Soria *et al.* (1975)		
Bean–sunflower	Fleck *et al.* (1984)		
Flax–wheat		Arny *et al.* (1929)	
Flax–oats		Arny *et al.* (1929)	
Sorghum–pigeonpea	Shetty & Rao (1981)	Shetty & Rao (1981)	
Pearl millet–peanut		Shetty & Rao (1981)	

(From Liebman, 1986.)

third species. The framework thus suggested might be useful in understanding the role of intercropping in weed control, and thus forms the basis of the first section of this chapter.

On the other hand, since the problem implicitly involves a positive modification of a piece of the environment of one species (the principle species) by another (the secondary species), especially, but not exclusively, in the case of a cover crop, it most obviously represents a case of the facilitative production principle. This formulation, leading to a different category of insights, is the basis of the second section of the chapter.

Competitive interactions among three species

The analysis of Levine (1976) provides the basic framework for this section. Consider two crops and a weed in various potential combinations, where crop 1 is C_1, crop 2 is C_2, and the weed is W. The weed (W) has a negative effect on both C_1 and C_2, both of which have a reciprocal negative effect on W. Thus

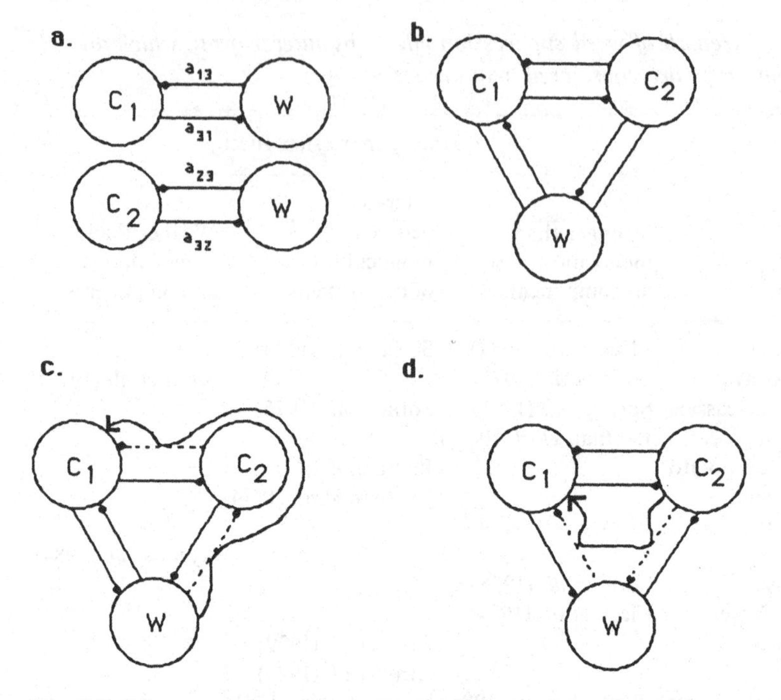

Fig. 8.2. Diagrammatic representation of three-species competition (*a*) two weedy monocultures; (*b*) a weedy intercrop; (*c*) illustration of the indirect positive effect of the weed on crop number 1; (*d*) illustration of the indirect positive effect of the second crop species on the first.

there is normal weed interference in either monoculture ($C_1 - W$, and $C_2 - W$). When the two crops are grown as an intercrop, we have the classic case of three species competition (see Figure 8.2). In addition to the crop to crop direct competitive effect and the weed to crop direct competitive effect, there also must be a crop to crop indirect effect through the weed. That is, the direct effect ($C_1 - C_2$) must be compared with the indirect effect ($C_1 - W - C_2$) if we are to know the true effect of C_1 on C_2 (and vice-versa). A simple application of Levine's analysis shows that (1) C_2 may be indirectly facilitated by the presence of C_1 (if the effect of C_1 on C_2 through the weed is greater than the effect of C_1 directly on C_2), or (2) C_2 may be facilitated by the presence of C_1 (same mechanism), or (3) both may be facilitating each other, in which case we have an indirect mutualism (Vandermeer, 1980*a*; Vandermeer *et al.*, 1985). In all cases above, the intercrop must be judged as advantageous given the usual LER criterion, since the relative yield of one or the other or both crops will be greater than unity.

These ideas are reviewed graphically in Figure 8.2. Figure 8.2(*a*) pictures the simple joint interaction between the crops and the weed, under the conditions of monocultural production, while Figure 8.2(*b*) illustrates the parallel situation of intercropping. The interactions illustrated are the pairwise interactions, that is, the direct effect of one plant on the other (in Figure 8.2(*a*) and (*b*)). But these direct effects are not the only ones of concern. The indirect effects may also be extremely important. While Levine's analysis was originally based on an assumption of equilibrium, which we naturally do not embrace here, the qualitative conclusions with regard to interactive effects are nonetheless potentially applicable to the intercropping situation. Suppose that the two intercropped species affect one another with an intensity equal to a_{12} and a_{21} (see Figure 8.2(*b*)), while the weed (call it species 3) exerts an effect a_{13} and a_{23} on the first and second crop species respectively. Finally, crop 1 and crop 2 exert an effect on the weed a_{31} and a_{32}, respectively. Thus, the effect of the weed on crop species 2 is a_{23} directly. But there is also an indirect effect. The weed also affects the first crop species, which is a competitor with the second. So the indirect effect, $a_{12}a_{23}$, is also part of the ultimate effect of the weed on crop species 1 (see Figure 8.2(*c*)).

This idea has particular importance when examining the competitive effect of the crops on one another. Suppose the crop of interest is C_1 and the monoculture is thus stipulated as $C_1 - W$. The unencumbered yield of the crop will be reduced by some fraction of the biomass of the weed, which we will call a_{13}. Adding the second crop species C_2, we add the direct effect a_{12}, and thus potentially reduce further the unencumbered yield of C_1, but also introduce a reduction of the weed through the competitive effect of the second crop species, a_{32}. The second crop thus exerts an 'indirect effect' which can be symbolized as $a_{13}a_{32}$. We picture this relationship in Figure 8.2(*d*). Generally, if $a_{13}a_{32}$ is larger than a_{12}, the second crop species facilitates the first through weed suppression (since the positive indirect effect in absolute magnitude is greater than the negative direct effect, making the overall effect positive).

This relationship can be portrayed more accurately by referring to the linear equations developed in Chapter 3 and adding a weed as a third competitor in the system. Thus, for the monoculture, representing the yield of C_1 as y_1,

$$\left.\begin{aligned} y_1 &= K_1 - a_{13}w, \\ w &= K_3 - a_{31}y_1, \end{aligned}\right\} \tag{8.1}$$

where K_1 refers to the crop yield in a completely weed-free monoculture, K_3 the weed biomass in a noncropped field, and the associated competitive effects are symbolized as a_{ij}, the competitive effect of the *j*th on the *i*th species (the weed is here treated as the third species, making a_{13} the competitive effect of the weed on crop 1).

The intercrop system, then, is given as

$$\left.\begin{aligned} y_1' &= K_1 - a_{13}w' - a_{12}y_2, \\ w' &= K_3 - a_{31}y_1' - a_{32}y_2, \\ y_2' &= K_2 - a_{21}y_1' - a_{23}w'. \end{aligned}\right\} \quad (8.2)$$

Using equations (8.1) and (8.2) we now ask under what conditions will y_1' be greater than y_1. The result is that if $K_2 - K_3$ is equal to or less than zero, y_1' will be greater than y_1 as long as $a_{13}a_{32} > a_{12}$ (see Appendix A for the rather tedious demonstration of this fact), qualitatively equivalent to Levine's result as described earlier.

The above analysis actually treats an extreme situation, one in which the basic process of three-species competition indirectly generates a facilitative effect. But this situation is not the only one in which an intercrop advantage accrues from weed supression. In fact, there appears to be some confusion with regard to the meaning of the statement 'the intercrop confers a measure of weed control'. There is actually a problem with the way this problem is posed in the first place. Taken at face value, presumably we mean to say that 'some of the intercrop advantage that we observe would be lost if the competitive effect of the crops against the weeds were diminished'. This idea can be accurately phrased with reference to the above system of linear equations. Specifically, the intercrop can be said to have a weed-suppressive effect if LER declines as a_{31} and/or a_{32} (the crops' effects on the weeds) are decreased. We can thus perform the thought experiment of decreasing the competitive effect of the crops on the weed, and ask what will happen to the relative yields. Using equations (8.2) as a model, it is possible to demonstrate that the relative yield of the first species will increase if, and only if,

$$(K_2a_{13} - K_1a_{23}) > (\Delta a_{31}/\Delta a_{32})(a_{12}a_{23} - a_{13}) + (a_{13}a_{21} - a_{23}), \quad (8.3)$$

while the relative yield of the second species will increase if, and only if,

$$(K_1a_{23} - K_2a_{13}) > (\Delta a_{32}/\Delta a_{31})(a_{13}a_{21} - a_{23}) + (a_{12}a_{23} - a_{13}), \quad (8.4)$$

where Δa_{31} is the proposed change in the effect exerted by the ith species on the weed (the derivation of equation (8.3) is presented in Appendix B). If neither (8.2) nor (8.3) are true, the increased competitive effect on the weed will decrease the LER. It is important to note that all of the competitive effects are involved in determining what will be the response of LER to increased (or decreased) weed control of either or both of the crops. It is thus not possible, in principle, to determine if an intercrop advantage is due to weed control simply by examining the LERs of a weedy versus non-weedy intercrop.

An alternative way of viewing this problem follows the same logic as nitrogen addition, as described in Chapter 5. Suppose it is possible to apply a weed-control technology at different intensities, such that one might go from no-control, through different levels of control, to complete control. Both

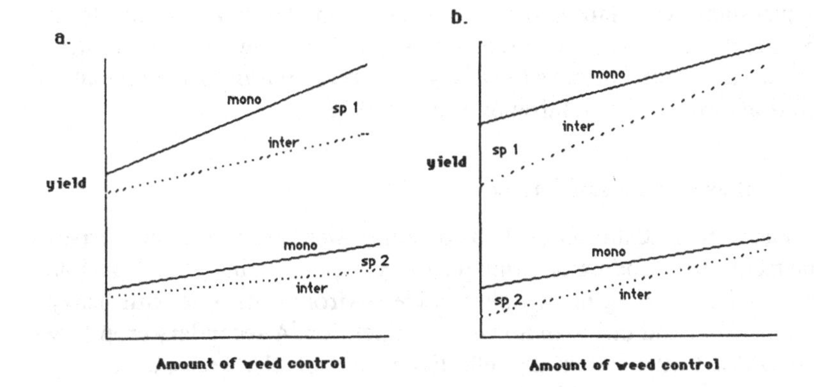

Fig. 8.3. Illustration of yield responses to weeding: (*a*) response that will give lower LER in a weeded intercrop than in a nonweeded one; (*b*) response that will give a higher LER in a weeded intercop than in a nonweeded one.

crops, in both monoculture and intercrop, are expected to increase yields as a function of intensity of weed control, whether or not the crops themselves have any effect on the weeds. This relationship is illustrated in Figure 8.3. Figure 8.3(*a*) and (*b*) shows two possible patterns, either of which could result independently of the weed-controlling effect of the crops. In Figure 8.3(*a*) the LER in the weedy intercrop is greater than in the weed-controlled intercrop while the reverse is true in Figure 8.3(*b*). Clearly, an examination of weedy vs nonweedy LERs provides us no information with regard to the weed-control mechanism of intercrop advantage.

This section has treated two distinct notions of intercrop advantage induced by weed control. First, we ask if one of the species (say the first) has a higher yield in a weedy intercrop than in a weedy monoculture. This will be true if the indirect positive effect of the second crop on the first ($a_{13}a_{32}$) is greater than its direct effect (a_{12}) (with the stipulation that $K_3 - K_2 \geq 0$, see Appendix A). Second, we ask whether the LER of a weedy intercrop will be increased if the competitive effect of the crops on the weed is increased. This will be true if equations (8.3) and (8.4) are true.

There is a danger in all of this analysis. While the qualitative conclusions are probably quite robust, a claim subject to further investigation, the specific form of the equations should not be taken too seriously. To assume linearity in the neighborhood of the intersection of the competition functions is probably not unreasonable. But to presume linearity over all possible yield values is suspect to say the least. One cannot, for example, compute competition coefficients from all pair-wise experiments and extrapolate to the set of three equations to predict the performance of a weedy intercrop (for example in making just such a calculation for a weedy intercrop of barley and peas (from data of Liebman, 1986), the predicted value of barley was a *negative* 85 gm/m^2). Plant competition is too likely to be very non-linear to make such computations worthwhile (Haizel & Harper, 1973).

Weed control as indirect facilitation

An alternative representation of the problem of weeds in intercrops derives from the facilitative production principle (see Chapter 4). Imagine a principle crop (say maize) existing in two alternative environments, one with heavy weed competition and one with no weed competition. A secondary crop (say beans) is added to the system and effectively controls the weeds, at least at some planting intensities. The basic idea is presented in Figure 8.4. As elaborated fully in Chapter 4, (and thus only sketched out here) the yields in each of the alternative environments can be presented as a potential set, as in Figure 8.4(*b*). The environments are then thought of as occurring as proportions of each, p percent of the time the environment will be weed-free. This is a convenient interpretation of the process because p then represents the relative effectiveness of the weed-control strategy employed by the producer. If weed control is 100% effective, $p = 1.0$, and the optimal intercrop is stipulated directly by the function that relates yield of the principal crop to the intensity of the secondary crop, in the no-weed environment. But if, for example, weed control is only 75% effective, the relevant combination of environments is 75% no-weed and 25% weed, employing the abstraction of thinking of an intensity as proportions of two extremes, as fully explained in Chapter 4. All the machinery previously developed for dealing with facilitative environmental modification can then be employed.

Specifically, Figure 8.5 shows the various possible qualitative situations that may exist. Figure 8.5(*a*) presents the situation of a convex potential set (weak competition and sensitive environment) which in this case signifies weak competition against the principle crop from the secondary crop but strong competition against the weed from the secondary crop. As the proportions of the environments change, which is to say, as the effectiveness of weed control (other than that from the secondary crop) increases, the optimal solution gradually shifts from the full intercrop to a smaller intensity of intercrop and eventually becomes the monoculture (when the effectiveness of control is 100%). It would be possible, with the proper information, to construct the potential set for a particular weed-crop situation and thus be able to predict the optimal secondary crop intensity. It is simply a matter of finding the point at which the adaptive function is tangent to the potential set.

The other extreme is a concave potential set, when the secondary crop is a strong competitor against the crop but a weak competitor against the weed. In

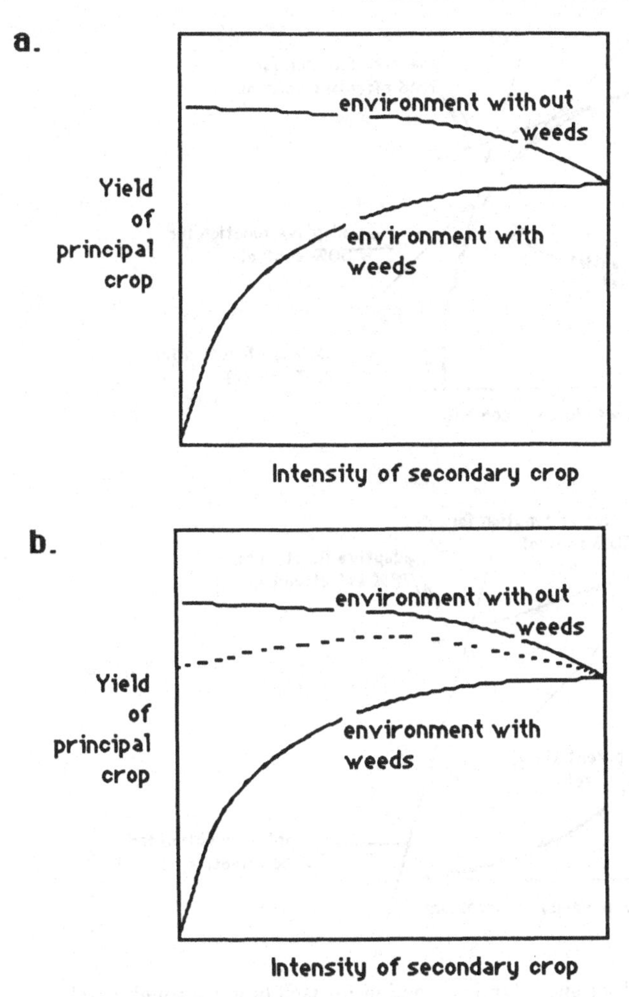

Fig. 8.4. Facilitation of principal crop by a secondary crop through weed control: (*a*) expected yields of principal crop as a function of intensity of secondary crop in the two extreme environments; (*b*) same as (*a*) with the intermediate environmental conditions (dotted line) added.

this case, illustrated in Figure 8.4(*b*), the optimal situation will be either the monoculture or a full intercrop, depending on the efficiency of weed control. With a relatively high efficiency, the monoculture is the optimal, and remains so with decreasing levels of efficiency, up to a critical level, at which the optimal solution switches to the full intercrop.

All of the above, however, ignores a critical practical point. Under

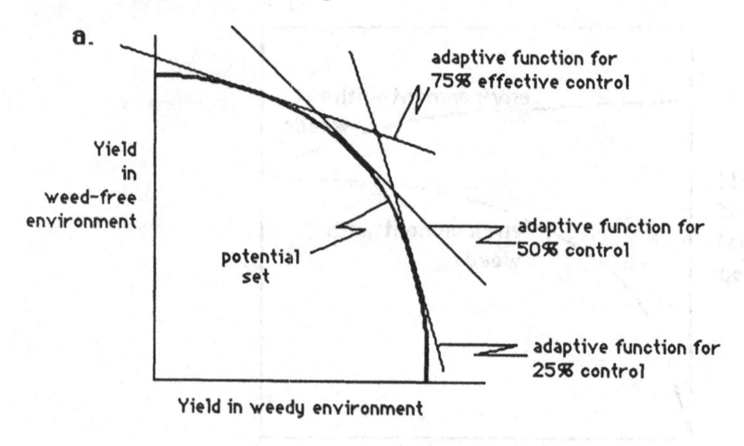

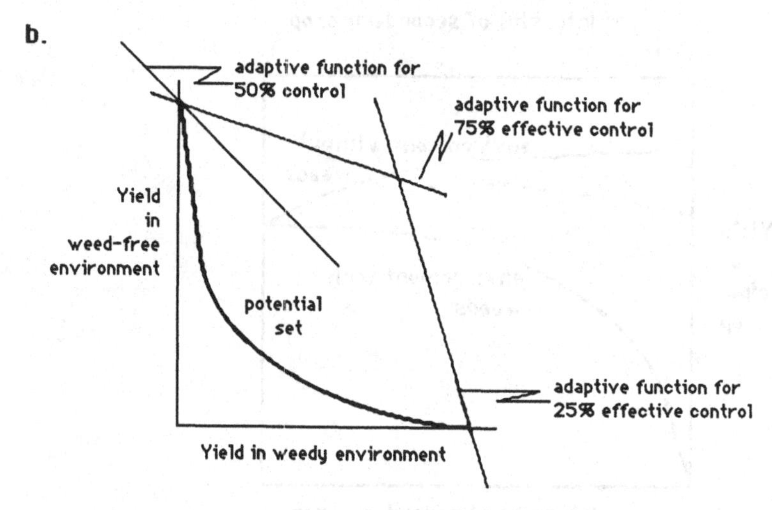

Fig. 8.5. The potential set and adaptive function for facilitation through weed control: (*a*) a convex potential set; (*b*) a concave potential set.

productive situations there will be a cost associated with maintaining a weed-free environment, whether it be the cost of herbicides or the cost of labor or machinery. Indeed, this cost is frequently cited by small producers as a major reason for intercropping (Norman, 1977; Olukosi, 1976; Lamberts, 1980), and several studies have demonstrated the main benefit of intercropping to be lowered weed-control costs (Ashokan *et al.*, 1985). The weed-control methodology may be effective, partially effective, or non-effective, but nevertheless has a cost. It is thus most appropriate to deal with the net yield (monetary returns discounted for cost of weed control) rather than the gross yield. It is

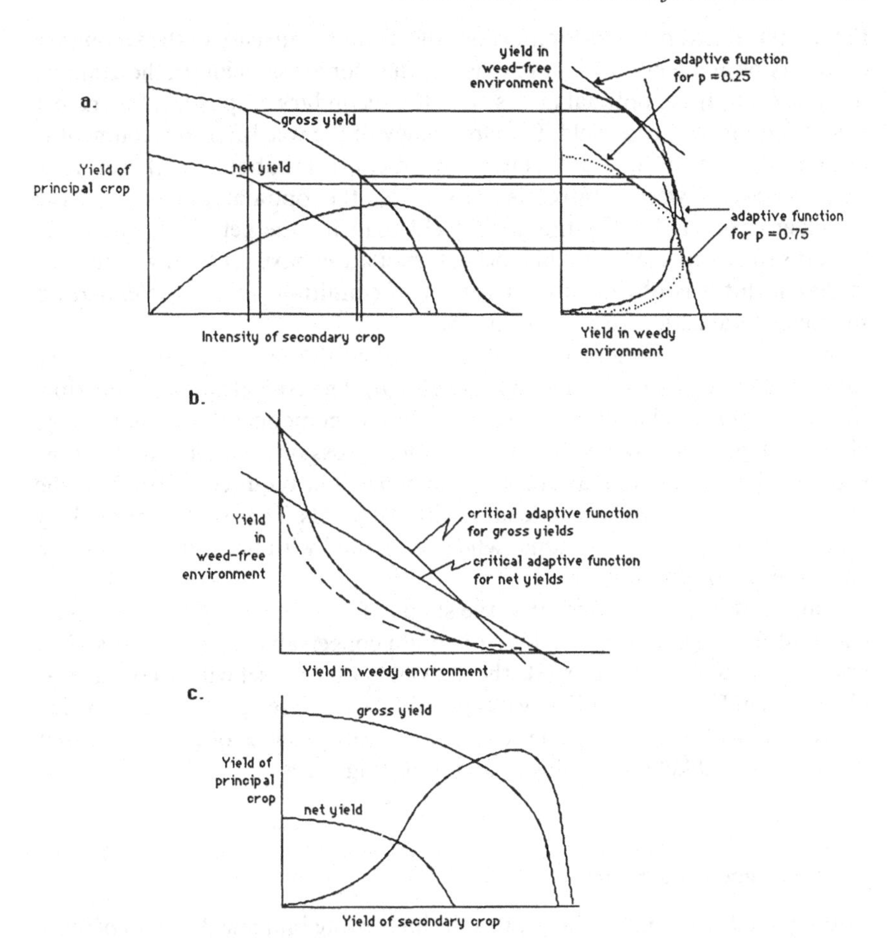

Fig. 8.6. Construction and analysis of potential set for gross and net yields (*a*) construction of the potential set; (*b*) maximising yield on concave potential set; (*c*) illustration of different shaped potential sets depending on whether yields are gross or net. In this case the gross yield will give a convex set and the net yield a concave set.

especially interesting to compare the results of dealing with gross yields to those that derive from using the net yields.

In Figure 8.6(*a*) I have illustrated the difference between net and gross yields, using the methodology employed in this section. Note that the potential sets may or may not intersect (the particular example in Figure 8.6(*a*) shows them intersected) but, most important, the optimal intensity of the potential sets may or may not intersect (the particular example in Figure 8.6(*a*) shows them intersected) but, most important, the optimal intensity of the secondary crop may be greater or lesser for net yields than for gross yields. In the example in Figure 8.6(*a*) the optimal intensity of the secondary crop is less for the net yield than for the gross yield if the frequency of the weed-free environment is around 0.75 (i.e., if the effectiveness of weed control is about 75%). But if the effectiveness of weed control is only 25%, the optimal intensity of the secondary crop is less for the gross yield than for the net yield. The exact patterns to be expected with this sort of formulation have yet to be worked out in detail. But it is obvious that a variety of conditions are possible, and no simple generalizations can be easily made.

A further interesting result is obtained when the potential set is concave rather than convex, as pictured in Figure 8.5(*b*). The critical adaptive function, that value of p at which the optimal switches from monoculture to intercrop, changes depending on whether one considers gross or net yields. It will most generally be the case that a smaller value of p will be required in order for the intercrop to be optimal when dealing with net yields. That is, it is more likely that an intercrop will be optimal when concerned with net yields than when concerned with gross yields.

Finally, it is not necessary that the shape of the potential function remain constant from gross to net yields. It is quite conceivable that the gross yield potential set will be convex while the net yield potential set will be concave, as illustrated in Figure 8.6(*c*). While it is possible to conjure up examples in which the reverse will be true, judging from the probable shapes of the curves, that seems to be unlikely for biologically interesting cases.

Summary and conclusions

Adding weeds to an intercrop creates the interesting biological system of three mutually competitive species. The most obvious method of analysis is to compare direct and indirect effects. Thus one can determine, at least theoretically, from first principles whether or not one crop has a facilitative effect on the second crop. If such a facilitative effect operates, it is appropriate to analyze the system as an example of the facilitative production principle, using the tools developed in Chapter 4.

But there is another sense, far more subtle, in which weed control confers an advantage on the intercrop. Using the LER criterion as a measure of intercrop advantage, one can define weed control as a mechanism if a decrease in LER is expected as a result of reducing the competitive effect of one or both crops on the weed. This formulation of the problem makes it rather inaccessible to typical weedy vs nonweedy LER computations. A higher LER with or without weeds is irrelevant to the mechanism of weed control in such a formulation, calling into question the usual experimental methodology of comparing LERs with and without weeds.

If competition against weeds is so severe as to produce a facilitative effect, there is no question that a weed-controlling mechanism is involved. But short of facilitation, weed control may contribute to intercrop advantage in a way that is not detectable through normal experimentation. As discussed in Chapter 4, it may be that alternative methodologies, such as target experiments, will be required to sort out this somewhat complicated problem.

Appendix A

The linear competition equations can be written

$$y_1 = K_1 - a_{13}w,$$
$$w = K_3 - a_{31}y_1,$$

for the two monocultures, and

$$y_1' = K_1 - a_{12}y'_2 - a_{13}w',$$
$$y_2' = K_2 - a_{21}y_1' - a_{23}w',$$
$$w' = K_3 - a_{31}y_1' - a_{32}y_2'.$$

Applying Cramer's rule, the solution for y_1' is

$$y_1' = (K_1 - K_1a_{23}a_{32} - K_2a_{12} + K_3a_{12}a_{23} + K_2a_{13}a_{32} - K_3a_{13})/\det \mathbf{A}, \quad (8A.1)$$

where **A** is the detached coefficient matrix of the above set of three equations. The solution for the weedy monocultures (the pair of linear equations describing y_1 and w) is

$$y_1 = (K_1 - a_{13}K_3)/(1 - a_{13}a_{31}). \quad (8A.2)$$

We seek conditions that specify

$$y_1' > y_1.$$

Substituting equations (8A.1) and (8A.2) and simplifying, we obtain

$$a_{13}a_{32} > a_{12},$$

and

$$a_{23}a_{31}a_{13}K_1(K_2 - K_3) \leq 0,$$

as the two conditions that stipulate $y_1' > y_1$. The derivation for y_2 is similar.

Appendix B

Using the three linear equations described in Appendix A, we can write the relative yield of species 1 as

$$RY_1 = \frac{\begin{vmatrix} K_1 a_{12} a_{13} \\ K_2 1 \ \ a_{23} \\ K_3 a_{32} 1 \end{vmatrix}}{K_1 \det \mathbf{A}} \tag{8B.1}$$

We seek those conditions in which the RY will increase if we increase the competitive effect on the weeds, a_{31} and/or a_{32}. Define a'_{31} and a'_{32} as the new, larger, values of the competition coefficients. We thus have

$$RY'_1 = \frac{\begin{vmatrix} K_1 a_{12} a_{13} \\ K_2 1 \ \ a_{23} \\ K_3 a'_{32} 1 \end{vmatrix}}{K_1 (\det \mathbf{A})^*} \tag{8B.2}$$

where (det **A**)* refers to the calculation of the determinant using the new values of the competition coefficients.

The condition for demonstrating an intercrop advantage due to weed control is taken as

$$RY'_1 > RY_1. \tag{8B.3}$$

Substituting (8B.1) and (8B.2) into (8B.3), and rearranging, we obtain

$$(K_2 a_{13} - K_1 a_{23}) > (\Delta a_{31}/\Delta a_{32})(a_{12} a_{23} - a_{13}) + (a_{13} a_{21} - a_{23}),$$

where $\Delta a_{3i} = (a'_{3i} - a_{3i})$.

9

Variability and intercrops

A frequently claimed advantage of intercropping is its capability of dealing with environmental variability, implicitly equivalent to the avoidance of risk (Abalu, 1977; Francis & Sanders, 1978; Reddy & Willey, 1985; Reich & Atkins, 1970). While it is common for intercropping reviews to contain sections on variability and risk (e.g. Aiyer, 1949; Kass, 1978; Mead & Riley, 1981; Norman, 1974; Willey, 1979*a*; Lamberts, 1980), only rarely is the subject a central focus. Three notable exceptions are contained in the work of Rao & Willey (1980), Pearce & Edmondson (1982), and Schultz (1984). The latter work specifically treats the first two, and forms the basis of the first two sections of this chapter (measurement and evaluation, and variability under competition and facilitation).

The ecological literature on variability, and/or stability, is enormous. Useful reviews can be found in several places (e.g. Goodman, 1975; May, 1972; McNaughton, 1977; Murdoch, 1975). It had generally been held that diverse systems are more stable, or less variable. When one component either flushes or comes close to extinction, it is more likely that another component will compensate for it if a number of components are available to do so, which would be more likely in a highly diverse system than in a more monotonous system. It was, and still is, a common-sense notion, which is why May's (1972) claim of the reverse was such a surprise. All things equal, a more diverse system is expected to be less stable, not more. While May's definition of stability may have been too restrictive for many ecologists, his conclusion was nevertheless a stimulating one, and led to much intense discussion and debate.

With this background of ecological theory, it is tempting to find an analogy to intercropping systems. Two species growing together is indeed a more diverse system than either one of them growing alone. But most of the theoretical formulations from ecology deal implicitly with questions of great diversity, tens or hundreds of species, and the sorts of qualitative and quantitative conclusions they draw are not likely to bear much fruit for the problem as applied to intercropping systems. On the other hand, as suggested by Schultz (1984), the approaches taken by agricultural scientists in the analysis of stability may be a fruitful source of ideas for the ecologist, a topic not appropriate for further discussion here.

Apart from the important question of measurement and evaluation, variability and risk can be viewed theoretically from two distinct vantage points. First, the change in variability as we move from a monoculture to an intercrop can be analyzed as a simple consequence of interspecific competition or facilitation, the approach implicitly taken in most of the literature (e.g. Rao & Willey, 1980; Rao *et al.*, 1981; Faris *et al.*, 1983; Schultz, 1984). The source of the variation is simply taken as the background environment applied equivalently to monoculture and intercrop alike.

Second, the problem can be analyzed from the point of view of the structure of the environment (Vandermeer, 1984*b*). This approach uses the vehicle of the potential set and adaptive function as described in Chapter 4, and is of interest in those cases where facilitation occurs.

In what follows, we first consider some special problems in the evaluation of yield variability (or what is frequently referred to in the agronomic literature as yield stability). This is followed by a general theoretical treatment of yield variability under competitive and facilitative production. Finally we approach the topic from the point of view of the structure of the environment.

Problems of measurement and evaluation

The measurement of intercrop advantage with regard to variability is a more complicated subject than that treated in Chapter 2. In intercrops there are actually five variables that must be taken into account, monoculture yield of crop 1, monoculture yield of crop 2, yield in intercrop of crop 1, yield in intercrop of crop 2, and combined intercrop yield. Each of these variables has an associated variance, and the interest is in two different forms of comparisons. First, the variability of crop 1 in an intercropping situation is compared to what it is in monoculture, and second, the variability of the combination in intercrop (the 'system' variability) is compared to what it is in the system of both monocultures. Both of these comparisons are interesting generally, and both speak to the question of variability in intercropping. Yet they are independent questions, and can easily have opposite answers.

Before dealing with this question in more detail, it is necessary to digress somewhat as regards the rationale of studying the variability, its measurement, and the basis of intercrop vs monocultural comparisons. With respect to the practical question, the real interest is in the avoidance of risk. Thus the most direct measure is the probability of falling below some prespecified level, for either of the monocultures, or either component of the intercrop, or the combination of components in the intercrop, as discussed by Rao & Willey (1980). But '. . . a theoretical understanding of risk itself must separate the contributions of mean, variance, and higher moments in yields to the probability of disastrous results'. (Schultz, 1984.) While the direct study of risk seems possible only through an *a posteriori* analysis of actual data, its theoretical study includes at least questions of the mean and variance of yields.

The standard measurement of variability is the coefficient of variation (the standard deviation as a proportion of the mean), since most applications are mainly concerned with the relative variability (e.g. corn varying between 1000 and 2000 with a mean of 1500 is not really more variable than sesame varying between 1 and 2 with a mean of 1.5, even though the range of variation in one case is 1000 and in the other only 1.0). It may be more accurate in some sense to use the standard deviation of the logarithms (Lewontin, 1966), although this has the apparent drawback of altering the shape of the distributions (Schultz, 1984). More important, the coefficient of variation is simpler and more readily amenable to analytical manipulation. Finally an alternative technique based on the regression of the yields against a so-called measurement of environmental quality (actually the deviation of the yield itself from the grand mean), used specifically by Rao & Willey (1980), will not be discussed here.

As to exactly what comparisons should be made when asking questions about relative variability, Rao & Willey (1980) made the first significant stride forward. They compared the variance of the sum of the intercrop yields to the variance of the sum of the monoculture yields, where the monocultures are taken as making up exactly the same proportion of the total as was obtained in the intercrop. So, in their example, they assumed that the monocultures were made up of 61% sorghum and 39% pigeonpea, thus making the yields correspond exactly in both monoculture and intercrop (i.e. the proportions of the total yield attributable to sorghum was the same in the intercrop and in the monoculture). Using this criterion they found that the intercrop was less variable than the monocultures.

Schultz (1984) made the simple observation that in the same sense that we must use optimal monoculture yields when computing land equivalent ratios (see Chapter 2), we really ought to use optimal monoculture combinations when evaluating the stability (variability) of an intercrop, rather than fixing the monocultures at arbitrary proportions. That is, if the goal is to determine whether the intercrop is less variable than monocultural alternatives, we ought to use those monocultures that show, as monocultures, the lowest variability possible. Let us suppose that of the available land, a fraction p is planted with crop 1 and a fraction $1-p$ is planted with crop 2. The problem is to determine what exact p will give the minimum coefficient of variation. The theoretical relationship between p and the coefficient of variation is plotted in Figure 9.1 for the original data of Rao & Willey. As can easily be seen, the fraction of land planted to sorghum which would give the minimal coefficient of variation is about 20%, a much smaller value than the 61% figure originally used. Also in

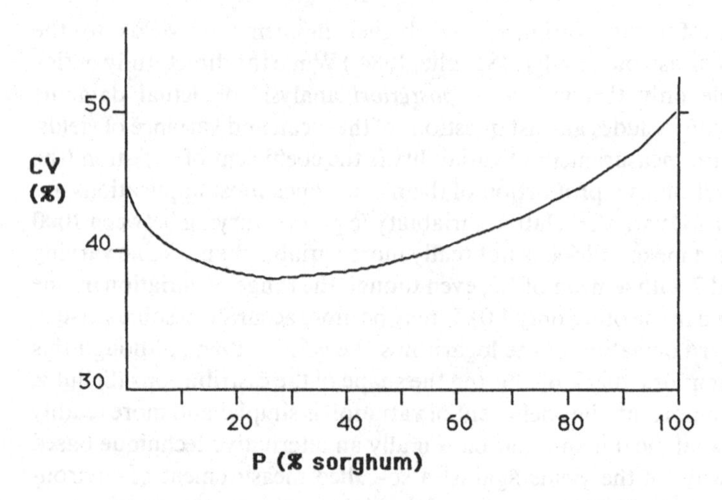

Fig. 9.1. The relationship between coefficient of variation and proportion of sorghum for the data of Rao & Willey (1980) (redrawn from Schultz, 1984).

Figure 9.1 are plotted the actual values of the coefficient of variation for the intercrop, showing that the intercrop is only apparently less variable than the monocultures, if one allows the comparison with the minimally variable monocultural combination. It is also true that some other intercrop might be less variable than the one actually studied. Indeed, if we were to require that both the intercrop and monocultures are those that give minimal variance, we must insist that even in this most carefully documented case, the question of whether or not the intercrop is more stable (less variable) is yet undecided.

Appendix A of this chapter presents the full mathematical treatment of how one goes about computing the monocultural combinations which give the minimal coefficient of variation for the monoculture yields.

Variability, competition and facilitation

While the evaluation of yield stability as an empirical problem is quite complicated and as yet unresolved, as described above, nevertheless a great deal can be said theoretically about how variability is generated and what might be the likely patterns as a function of various biological and cultural forces. In this section we explore the necessary consequences of competitive and facilitative production in terms of variability. More details of these developments can be found in Schultz (1984).

The yields of any monoculture will vary over time and over space. Let us for the moment ignore whether the variation is over time or space, and simply ask

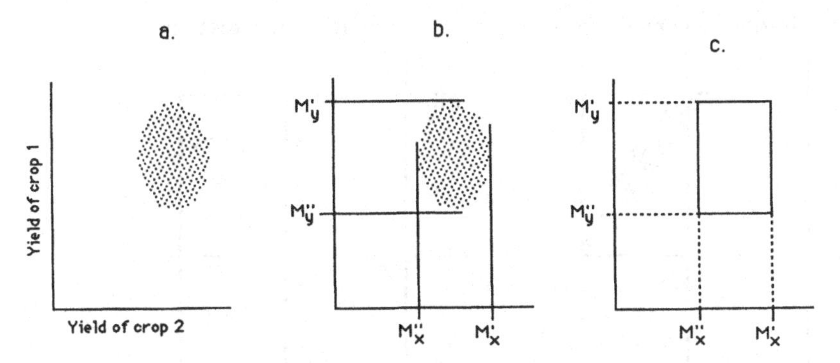

Fig. 9.2. Illustration of intercrop variability and the construction of the variability 'envelope': (*a*) the cloud of points in an intercrop system; (*b*) The extreme values of the cloud of points on each axis; (*c*) the envelope, formed from the extreme values, for the cloud of points.

how would that variation appear if we had two crops and plotted their yields on a graph of one against the other. Such a representation is provided in Figure 9.2(*a*). A cloud of points will be observed, a large cloud if the variability is relatively large and a small cloud if the variability is relatively small. For purposes of the qualitative developments sought here, it is convenient to think of the magnitude of the variability as being directly proportional to the magnitude of the cloud of points.

It is a simple matter to identify the extreme values along each axis, as has been done in Figure 9.2(*b*). In this example we can expect the yield values for the crop on the x axis to be somewhere between M_x' and M_x'', while the yield values of the crop on the y axis will be somewhere between M_y' and M_y''. These extreme values can thus be thought of as defining a rectangular area within which the cloud of points will be found. This area is called the monocultural envelope, and simply refers to the rectangular outer limits of the actual yields of the monocultures, as pictured in Figure 9.2(*c*).

Note that the cloud of points need not be symmetrical within the envelope. For example, in Figure 9.3 are two examples of nonsymmetrical clouds, within the same envelope. If the underlying forces causing the variability affect the two crops in a similar way, the expectation is that the two will vary more or less in the same way. Thus there will be a general positive correlation between their yields, as pictured in Figure 9.3(*b*). But it is equally plausible to suggest that the underlying forces have opposite effects on the two crops, thus generating a negative correlation, as depicted in Figure 9.3(*a*). We will have cause to return to these correlations in a moment.

The cloud of points representing the monocultural yields (or the single point as it was represented in Chapter 3) will be reduced on both axes through

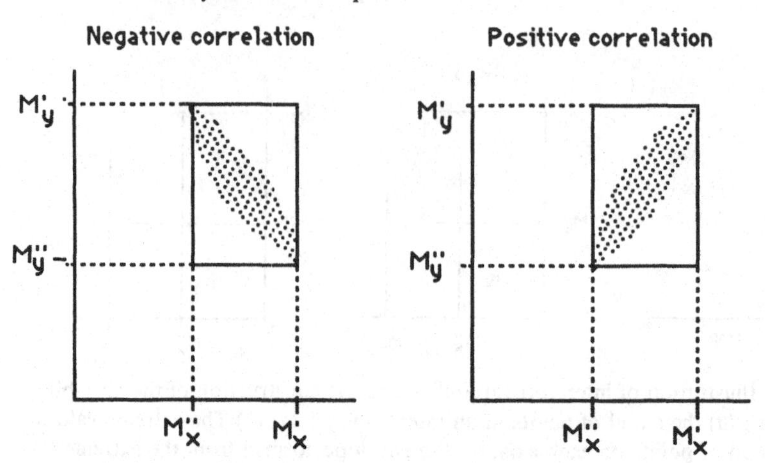

Fig. 9.3. Effect of the correlation between crop yields on the position of the cloud of points within the envelope.

competition. Just as the expected yields were reduced on each axis as a consequence of competition, here the expected cloud of points will be reduced for the same reasons. But here the yields will be a cloud of points, created by the same underlying forces that created that variability in the monocultures. Thus, in Figure 9.4(*a*) we present the cloud of points for a tomato–cucumber system (Schultz *et al.*, 1982), both the monoculture and intercrop clouds. As in Chapter 3, the expected points are reduced according to the amount of competition experienced. But here, since we began with variability in the monocultures, that variability is reflected in the intercrop likewise. Thus it is necessary to translate the monoculture envelope from its original position to a position in which its center (that is the point representation of the two average monocultural yields) corresponds to the center of the intercrop cloud, retaining the same exact coefficient of variation that existed in the original monoculture envelope. A convenient method for accomplishing this transformation graphically is with the range and median of the cloud of points. We take the ratio between the range and median to be a measure of the relative variability of the system. It is thus a simple task to translate the monocultural envelope to a position in which its center is on the center of the intercrop cloud, reducing the size of the envelope such that the new envelope retains the relative variability of the original monoculture envelope. This new envelope we call the 'equivalent' monoculture envelope. It represents the limits on a cloud of points that have the identical relative variability as the monocultures, but with the mean values reduced to be equivalent to the intercrop yields. In Figure 9.4(*b*) the original monoculture envelope is scaled down so as to have its center on the intercrop means, retaining its initial relative variability. As can be seen, in

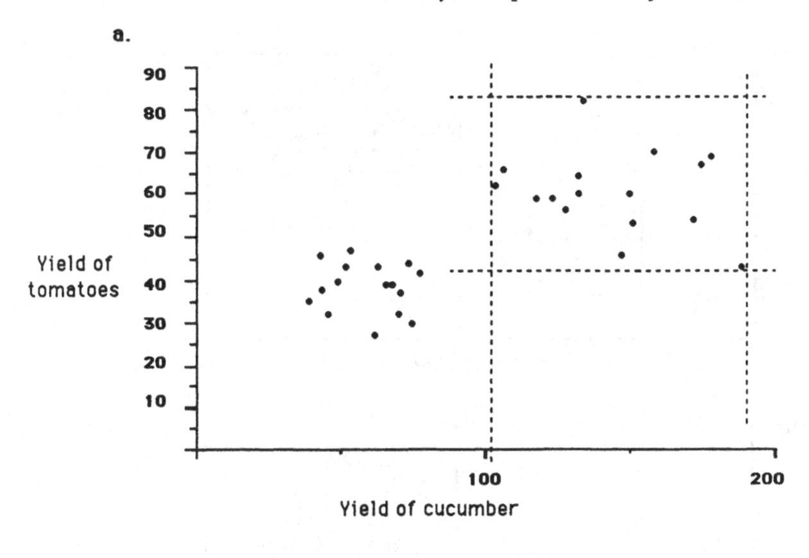

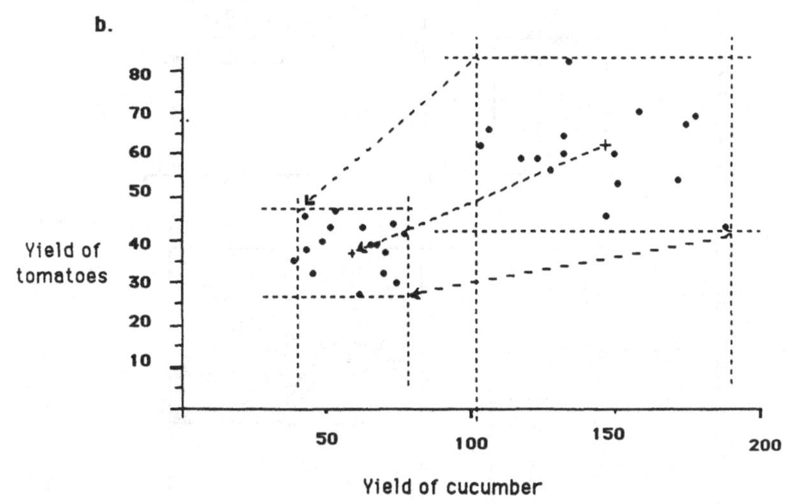

Fig. 9.4. Example of monoculture and intercrop envelopes. Points to the right are from monocultures of tomatoes and cucumbers, points to the left from the intercrop of the two (data from Schultz *et al.*, 1982): (*a*) the intercrop envelope; (*b*) translating the intercrop envelope to the monoculture position, retaining the same relative variation.

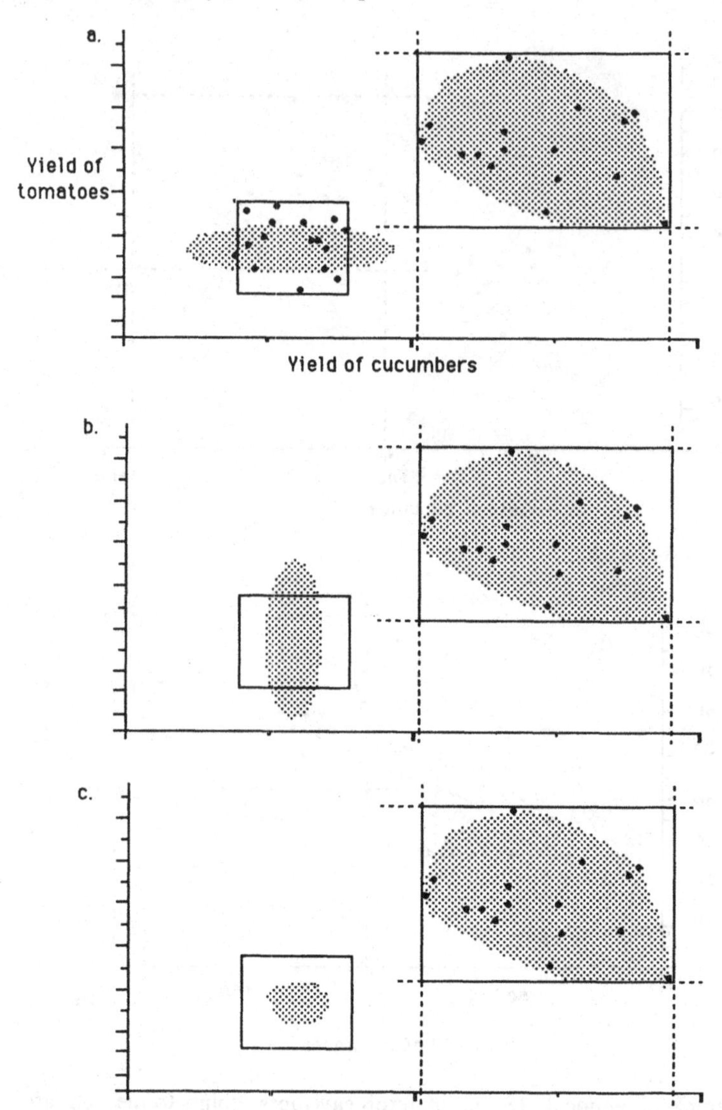

Fig. 9.5. Examples of possible variation patterns. Points are from a tomato-cucumber intercrop (Schultz *et al.*, 1982) and shaded areas indicate relative position of the cloud of points (actually for the monoculture, and theoretically for the intercrop).

this case the cloud of points representing the intercrop falls almost exactly inside the equivalent monocultural envelope. Had the intercrop cloud been different the conclusions could have been different. For example, Figure 9.5 presents theoretical examples where the intercrop cloud (the shaded area) is such that the cucumber variability is greater but the tomato variability less in the intercrop (Figure 9.5(*a*)), or the reverse (Figure 9.5(*b*)), or the total variability is less (Figure 9.5(*c*)). Note that this method is virtually identical to comparing the coefficients of variation between monoculture and intercrop. It has the advantage that it is easily visualized graphically and, what is most important here, facilitates a further qualitative theoretical development.

The next problem is to construct, theoretically, the cloud of points representing the intercrop yields. That cloud of points, as already described, must be less than (on both axes) the cloud of points representing the monocultural yields, presuming that the process of competition is operative. We can construct an envelope which will enclose the intercrop cloud with various models. For simplicity of presentation we presume that the interaction function is linear (see chapter 3), such that the yields of the two crops are related to one another as

$$\left.\begin{aligned} y &= M_y - ax, \\ x &= M_x - by, \end{aligned}\right\} \tag{9.1}$$

where, as in previous chapters, x and y are the yields in the intercropping situation, a and b are the competition coefficients, and M_y and M_x are the monocultural yields. If we presume that all of the environmental variability occurs in the monocultures, and that the interaction coefficients remain unvariable, we obtain the pattern shown in Figure 9.6(*b*). That is, the two functions are linear, and their variability is represented as a series of parallel lines, the extremes of which are shown in Figure 9.6(*a*). Under this set of assumptions, if we translate the monocultural envelopes so that its center is positioned at the center of the intercrop envelope, retaining its original coefficient of variation, the translated monocultural envelope will be exactly inscribed in the intercrop envelope, as shown in Figure 9.6(*b*).

With this representation we can deduce two very important facts. First, since the monocultural envelope is inscribed inside the intercrop envelope, it is not possible for the monoculture to be more variable than the intercrop, in terms of either the component intercrop variabilities, or the variability of the total yield. This conclusion is profoundly at odds with the conventional wisdom that intercrops are less variable than monocultures, suggesting not only that it is not logically necessary for intercrops to be less variable than monocultures, but that it is not even possible for them to be so. Second, as pictured in Figure 9.6, depending on the correlation between the crop yields, the intercrop can be either equivalent to the monocultures or greater than

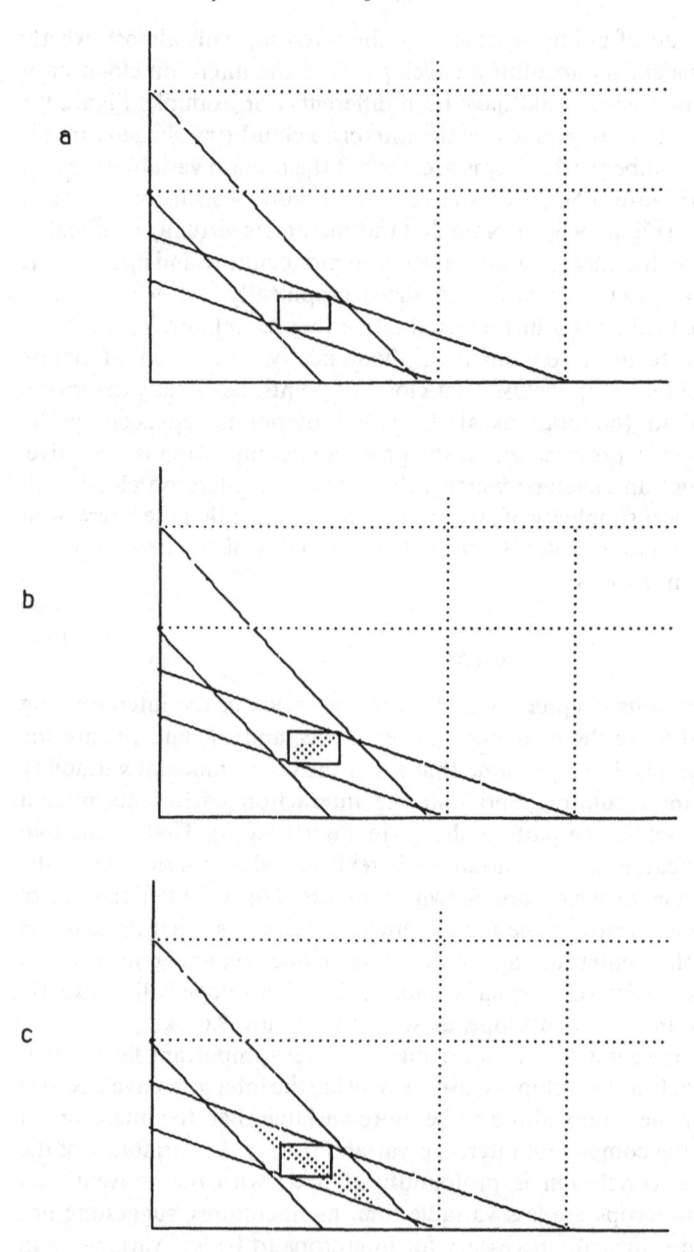

Fig. 9.6. Theoretical construction of intercrop cloud and its relationship to monocultural envelope (see text).

them, in terms of variability. Specifically, if the crops are positively correlated (Figure 9.6(*b*)) the variability of the monocultures (again either in terms of individual components or the total) and the variability of the intercrops are expected to be nearly identical. But if the crops are negatively correlated (Figure 9.6(*c*)), the intercrop is expected to be more variable than the monoculture.

The above analysis was based on the underlying assumption that the variability of both crops is equal in monoculture, the monocultural yields are identical, and that the competition coefficients are constant, in addition to the basic assumption of a linear competition function. We now relax those assumptions.

Let the monocultures of species i range from M_i to M_i', and the median be $M_i = (M_i + M_i')/2$. The mean intercrop yields are symbolized as y^* and x^*, and are given by the solution to equations (9.1), which are

$$\left.\begin{aligned} y^* &= (M_y - aM_x)/(1-ab), \\ x^* &= (M_x - bM_y)/(1-ab). \end{aligned}\right\} \quad (9.2)$$

The relative variability of the monocultures will be $(M_i' - M_i)/M_i$. Thus to retain the same relative variability for the intercrop we require

$$(M_x' - M_x)/M_x = R_x/x^*,$$

in which R_x is the length of the equivalent monoculture envelope along the x axis. Solving for R, using equation (9.2), we obtain

$$R_x = \frac{(M_x' - M_x)(M_x - bM_y)}{M_x(1-ab)},$$

and the corresponding equation for y. The area (and thus the joint relative variability) of the equivalent monoculture envelope is R_xR_y, or

$$V_m = R_xR_y = \frac{(M_x' - M_x)(M_x - bM_y)(M_y' - M_y)(M_y - aM_x)}{M_xM_y(1-ab)^2}.$$

Presuming the linear model as described above, we compute the area of the intercrop envelope, which we take as proportional to the intercrop variance, as

$$V_1 = \frac{(M_x' - M_x)(M_y' - M_y)}{1-ab}.$$

The intercrop relative variability will be smaller than that of the monoculture when $V_1 < V_m$, or, after simplification,

$$bM_y^2 + aM_x^2 < 0, \quad (9.3)$$

which is patently impossible as long as all the terms are positive. We are thus forced to at least the tentative conclusion that, as long as only competition is operative, intercrops will tend to be more variable than monocultures, both in terms of the variabilities of individual components and in terms of the variability of the combination.

Using equation (9.3) we can quickly extrapolate the exact opposite conclusion in the case of double facilitation (mutualism). If both a and b are negative (as they would be in the case of mutualism) it is not possible for the intercrop to have a larger relative variability than the monoculture, whereas in the case of single facilitation (a positive and b negative) the apportionment of relative variability will depend on the relative sizes of aM_x^2 and bM_y^2. We thus have the generally surprising result that intercrop relative variability will tend to be larger than that of the monocultures if competition is the main operative force, whereas intercrop relative variability will tend to be smaller when facilitation is the main operating force.

But these striking results are all based on the underlying assumption of the linear model, including the tacit assumption that the monocultural points will be relatively symmetrically shaped, as they are in Figures 9.4, 9.5 and 9.6. If the intercrop produces a correlation between the yields of the two species, the intercrop cloud of points is restricted to only a subset of the intercrop envelope. As shown in Figure 9.7, it is conceivable to obtain an intercrop relative variability that is smaller than that of the monoculture if the correlations between yields are positive. Such a possibility is especially likely when the relative variabilities of the two monocultures are very different. Schultz (1984) gives the exact conditions for such an occurrence, analyzing the coefficient of variation rather than this simplified graphical form.

Schultz (1984) presents several other modifications of the above arguments that can theoretically lead to intercrops having lower variances than the monocultures. If the competition coefficients are nonlinear and the yields are correlated with respect to the environmental variability (e.g. as in Figure 9.6(a)), it is possible to have a lower variance in the intercrop. Similarly, if the competition coefficients are positively correlated with the yield potentials as both vary through time, it is again possible but not necessary to find intercrop variability lower than that of the monocultures.

But even with these modifications, the general conclusion seems to be that under a competitive model, intercrops will tend to be more variable rather than less so, as stated by the conventional wisdom (but see Marshall & Brown, 1973). On the other hand, the exact opposite is true for a mutualistic model, with an intermediate situation for a single facilitation model. Since the conventional wisdom of reduced variability in intercrops does not seem to be supported by available evidence, nor is there a theoretical reason for expecting it to be generally true, it seems that a more reasonable null hypothesis would be that intercrops will tend to be more variable than monocultures if competition is operative, that they may be either more or less variable when facilitation is operative, and that they will tend to be less variable in those rare cases where mutualism operates.

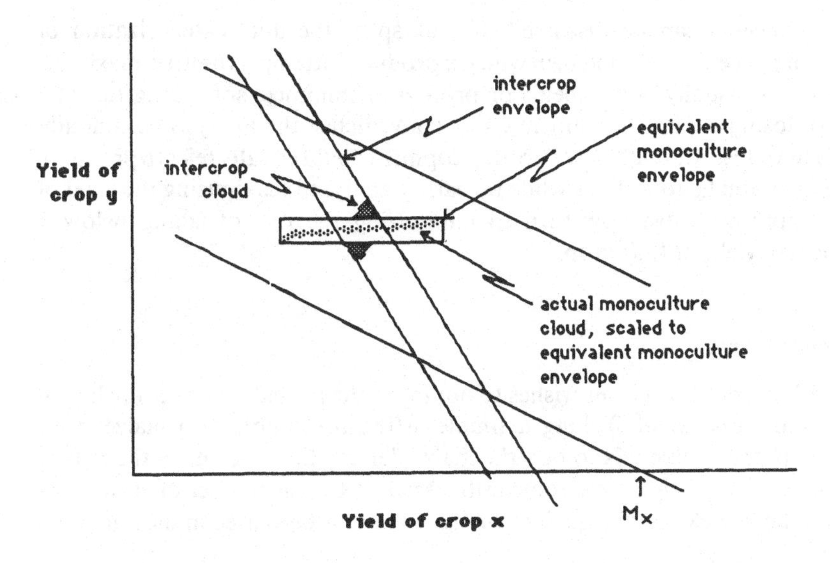

Fig. 9.7. Illustration of decreased intercrop variability (for crop x) due, in part, to large differences in variability along each axis.

Risk and facilitation

When facilitation is operative, the methodology introduced in Chapter 4 provides a useful tool of analysis. Beginning with a potential set we examine the maximization of yield on that set through the use of an adaptive function. (I presume that the reader is familiar with the methods introduced in Chapter 4.)

First consider the set of factors that causes a producer to choose one system or another. These factors are many and varied, but can be conveniently summarized in two broad categories, those factors that are sought (high yields, profits, etc.) and those factors that are avoided (yield below some critical level, a critical low value of marginal profit, etc.). For example, a typical midwestern U.S. farmer who chooses to grow grains is choosing to accept a relatively low, but assured profit. A neighboring farmer who grows tomatoes or pickles is choosing to accept a relatively high chance of losing everything, but also seeking the high profits that sometimes come with these crops. The first farmer seeks to avoid the risk, the second to maximize profits. These two relatively distinct strategies dictate (at least in part) the choice of cropping systems.

While simplistic categorizations such as these are always fraught with inconsistencies they nonetheless provide a convenient framework in which other concepts can be discussed. In this spirit the dual categorization of decision processes is adopted, in which a producer attempts to either maximize something (usually either yields or profits) or minimize something (usually risk of losing capital investment). For convenience the analysis continually refers to yields, although the entire development could equally refer to profits or other decision factors. Thus while we earlier analyzed maximizing the yield of a principal crop, we now turn to minimizing the risk of falling below a particular yield of that crop.

A simple example

The risk aversive producer wishes to minimize the probability of going below some particular yield. Taking a simple artificial example of a maize–bean intercrop, the problem is to find the maize density that minimizes the risk of the bean yield going below some critical value. Consider a set of exemplary data, of bean yields in two distinct environments, as presented in the following table:

Maize density	0	1	2	3	4	5
Yield in env. 1	28	27	24	21	13	12
Yield in env. 2	0	6	8	11	12	12

Now suppose there is random variation around each of those numbers, so that any yield might be either one or two units above or one or two units below the yield given in the table. So, for example, at maize density 1 in environment 1, we expect yields of 25, 26, 27, 28, or 29, with equal probability. The various expectations are given in the table opposite.

We now assume that the critical yield above which the producer must remain is 12, and thus ask, what is the probability of falling below 13? In the table opposite an asterisk is found next to those values that fall below 13. Assuming the two environments are coarse-grained and occur in equal proportions, since there are ten numbers altogether, one need only divide by 10 the number of times the yield will fall below 13 to obtain the probability. These probabilities are given in the penultimate row of the table. In the ultimate row are given the average yields. Note the optimum maize density is different for risk minimization (maize density 3) than for yield maximization (maize density 1).

Maize density	0	1	2	3	4	5
Range of yields in env. 1	26	25	22	19	11*	10*
	27	26	23	20	12*	11*
	28	27	24	21	13	12*
	29	28	25	22	14	13
	30	29	26	23	15	14
Range of yields in env. 2	0*	4*	6*	9*	10*	10*
	0*	5*	7*	10*	11*	11*
	0*	6*	8*	11*	12*	12*
	1*	7*	9*	12*	13	13
	2*	8*	10*	13	14	14
Prob. of going below 13	0.5	0.5	0.5	0.4	0.5	0.6
Average yield	14	16.5	16	16	12.5	12

The adaptive function under risk minimization

The first optimization criterion considered was yield optimization, in which the adaptive function was linear in both fine-grained and coarse-grained environments (Chapter 4). These results can be conveniently reviewed in Figures 4.8, 4.9(*a*), and 4.9(*b*). We now turn to the alternative decision criterion, risk minimization.

Consider first the fine-grain situation. Suppose that a minimum possible yield exists (referred to as min), probably zero in most practical situations, but here considered to be some positive value, for the sake of generality. Likewise, there is some maximal value (referred to as max), usually equal to the yield of the monoculture in the competitive environment, but here considered to be some positive value, again for the sake of generality. There will be some critical value of the yield, intermediate between the maximum and minimum. Yields below this critical value are unacceptable. The problem then is to compute the probability of avoiding the region below the critical value, over a long time period.

Assuming variability in the system, an expected yield which is close to the maximum is unlikely ever to produce a value below the critical value. An expected yield which is close to the minimum is very likely eventually to produce a value below the critical value (since the minimum is by definition below the critical value). Assuming that the relative position of the expected yield between max and min is equal to the probability of successfully avoiding the crash (i.e. falling below the critical value), we have, for the competitive portion of the environment,

$$\frac{Y_{\rm I}-\min}{\max}$$

and, for the facilitative portion of the environment,

$$\frac{Y_{\rm II}-\min}{\max},$$

as the probability of success in each of the two environments. Then the probability of success over the whole environment must be simply the average of the probability of occurrence of the two environments (by the definition of fine-grained), or

$$p_{\rm s}=p\left(\frac{Y_{\rm I}-\min}{\max}\right)+(1-p)\left(\frac{Y_{\rm I}-\min}{\max}\right).$$

Since this is the probability of success over one year, the probability of success in two years running will be $p_{\rm s}^{2}$, in three years running, $p_{\rm s}^{3}$, and in n years running, $p_{\rm s}^{n}$. Thus the probability of avoiding a crash for n years running, (p_n), is

$$p_n=\left[p\left(\frac{Y_{\rm I}-\min}{\max}\right)+(1-p)\left(\frac{Y_{\rm I}-\min}{\max}\right)\right]^n,$$

which can be altered algebraically to read

$$p_n^{1/n}\max+\min=pY_{\rm I}+(1-p)Y_{\rm II}.$$

If we allow $A=p_n^{1/n}\max+\min$, we have

$$A=pY_{\rm I}+(1-p)Y_{\rm II},$$

which is qualitatively identical to equation (4.1), the adaptive function in a fine-grained environment under yield maximization. Thus the optimization procedure in a fine-grained environment is essentially the same for either a yield-maximization procedure or a risk-minimization procedure. The various conclusions drawn for yield maximization in Chapter 4 are thus identical for risk minimization.

In a coarse-grained environment we are concerned with year-to-year or patch-to-patch probabilities, but the same probabilities as presented above. Thus, if year 1 is a competitive environment, the probability of avoiding the crash is $(Y_{\rm I}-\min)/\max$. In year 2 suppose again that the environment is competitive. The probability of avoiding the crash for both years is $[(Y_{\rm I}-\min)/\max]^2$. Now suppose that in the third year the environment is facilitative. The probability of avoiding the crash for three years running then is $[(Y_{\rm I}-\min)/\max]^2[(Y_{\rm II}-\min)/\max]$. In general, the probability of success for n years running will be

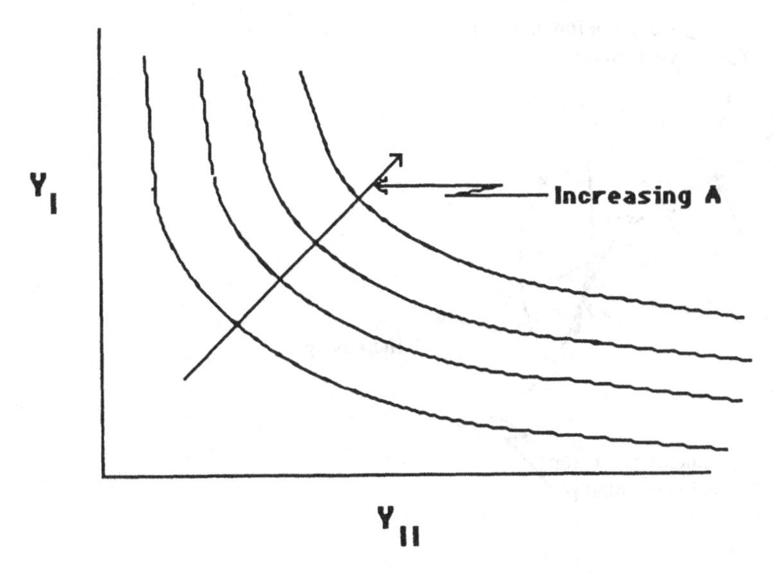

Fig. 9.8. The adaptive function under risk minimization in a coarse-grained environment (see text).

$$p_n = [(Y_{\rm I} - \min)/\max]^{pn}[Y_{\rm II} - \min)/\max]^{(1-p)n},$$

and the adaptive function is

$$A = p_n{}^{1/n} = \left[\frac{Y_{\rm I} - \min}{\max}\right]^p \left[\frac{Y_{\rm II} - \min}{\max}\right]^{(1-p)} \tag{9.4}$$

Equation (9.4) thus represents the adaptive function for risk aversion in a coarse-grained environment. Contrary to all previous cases it is not linear in $Y_{\rm I}$ and $Y_{\rm II}$, but rather follows a hyperbolic form, as illustrated in Figure 9.8.

The optimization problem

Simple optimization procedures on a convex potential set follow the same pattern as with a linear adaptive function, as pictured in Figure 9.9. When p is large (usually a competitive environment) the optimal strategy is the monoculture. As p decreases, the optimal strategy moves along the edge of the potential set, the optimal system slowly moving towards the position of full intercrop. Finally, when p becomes very small the optimal is the full intercrop.

Recalling the data from the artificial example in the table (p. 155), remember that the optimal solution was obviously different for risk aversion than for yield optimization. With the two different forms of adaptive function,

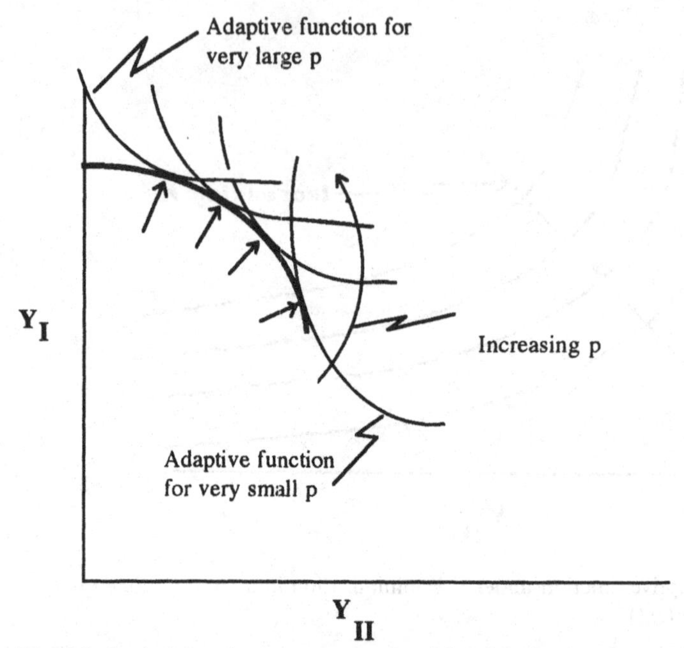

Fig. 9.9 Optimizing the intercrop under risk minimization for a convex potential set. From Vandermeer 1984b.

we see how these optima come about in Figure 9.10. The risk aversive adaptive function intersects the potential set at the point corresponding to a maize density of 3 while that of the yield maximizing function indicates a maize density of 1, just as was computed earlier from the numerical example.

To analyze the same situation with a concave potential set we make use of Levins' original approach (1968). As shown by Schultz (1984), it will sometimes be the case that the optimal system for the avoidance of risk will be a combination of monoculture and polyculture, the so-called 'mono-poly' solution. It is possible to visualize all possible combinations of systems as the *extended potential set* (equivalent to Levins' extended fitness set (Levins, 1968)). That is, suppose 50% of the cultivated area is planted with a monoculture and 50% with the full intercrop. The yield of the entire area then will be 50% that of the monoculture and 50% that of the full intercrop, graphically, a point half-way between the points of the monoculture and the full intercrop (Figure 9.11). All other possible combinations of full intercrop and monoculture can be represented on the line connecting the monoculture with the full intercrop (Figure 9.11).

If we thus seek to maximize the adaptive function, free from the restraint of

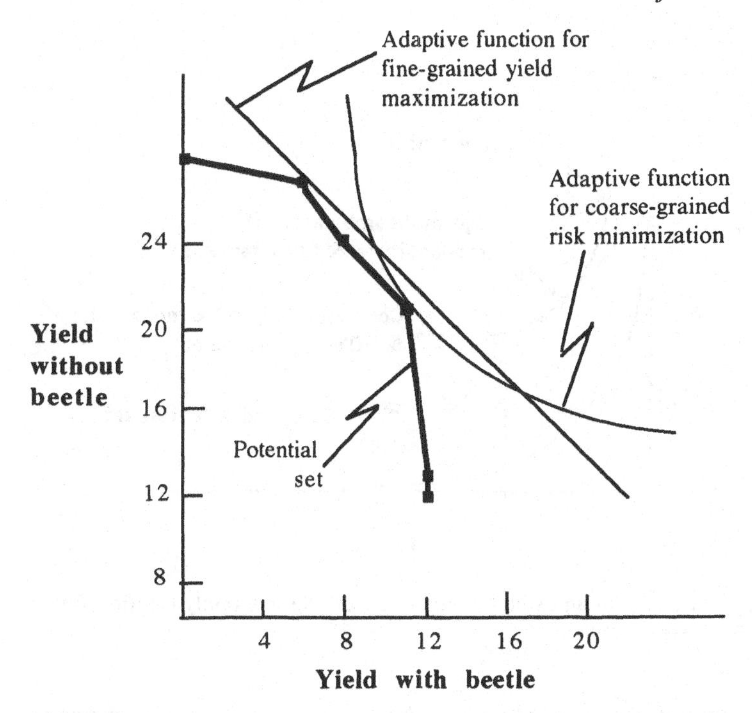

Fig. 9.10. Optimal solution for example in text. Optimization on potential set is different for the two different adaptive functions.

using only one system, we will find the largest value of the adaptive function which intersects the extended potential set. As can be seen from Figure 9.11, that solution will almost always be some mono-poly solution, that is, some combination of the two extremes. Thus, when the environmental grain is coarse and the producer operates under a risk minimization criterion, and the potential set is concave (an insensitive environmental factor and highly competitive secondary crop), the best solution is to plant some land in monoculture and some in intercrop.

This latter result is, on first glance, a surprising and nonintuitive one. The theory suggests that when a secondary crop is highly competitive, it should only be combined with the principal crop when it affords a very good facilitative environment. But if the facilitative environment occurs in an all or nothing fashion (i.e. is coarse-grained), it makes sense in the long run to plant some intercrops (for the years when the facilitative environment occurs), and some monocultures (for the years when the competitive environment occurs).

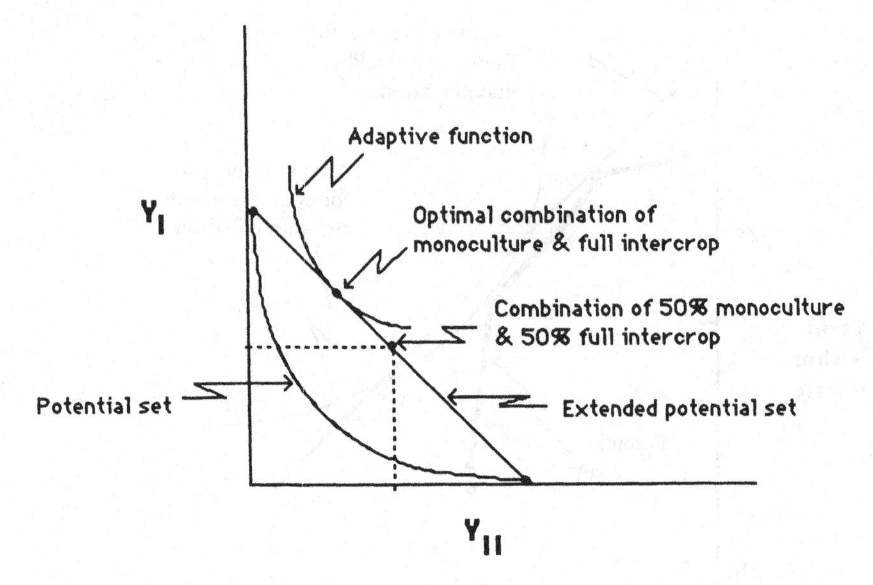

Fig. 9.11. Optimization on an extended potential set, the monopoly solution (see text). From Vandermeer 1984b.

Under these circumstances it is better in the long run to take a loss on one of the plantings but a large gain on the other, rather than taking a loss every *p*th year on the monoculture or accepting lowered yields $(1-p)$ of the time in the intercrop.

Appendix A

Following Schultz (1984) we note that for a given crop x and crop y it is true by definition that:

$$s^2_{x+y} = s^2_x + s^2_y + 2r_{x,y}s_x s_y, \tag{9A.1}$$

$$\mu_x \mu_y = \mu_x + \mu_y, \tag{9A.2}$$

$$s^2_{kx} = k^2 s^2_x : s_{kx} = ks_x, \tag{9A.3}$$

where s_{x+y} is the variance of the sums of sole crop yields for each site, $r_{x,y}$ the correlation between x and y, k is any given constant, and the subscripted s^2, s, and μ are the variances, standard deviations, and means respectively.

To find the least variable combination in terms of the coefficient of variation of two sole crops, we first weigh the yields obtained per unit area for each crop by the proportion of the total area planted so that we have px and consequently $(1-p)y$, where p is the proportion of the area planted in crop x, and $(1-p)$ is the proportion planted in crop y. We then use the equalities from equation (9A.3) to substitute p and $(1-p)$ into equations (9A.1) and (9A.2) and find the coefficient of variation of the total yields for a proportionate combination of sole crops, given a particular value of p. Thus we have

$$CV_{x+y} = \frac{(p^2 s^2_x + (1-p)s^2_y + 2p(1-p)r_{xy}s_x s_y)^{\frac{1}{2}}}{p\mu_x + (1-p)\mu_y}. \tag{9A.4}$$

Differentiating equation (9A.4) with respect to p, we set it equal to zero, and solve for p to obtain the proportionate area of a crop x, designated p^*, expected to give the most stable sole crop. This gives us

$$p^* = \frac{\mu_x s^2_y - \mu_y r_{x,y} s_x s_y}{\mu_y s^2_x + \mu_x s^2_y - (\mu_x + \mu_y) r_{x,y} s_x s_y}. \tag{9A.5}$$

Note that p^* may take on values less than zero or greater than one, which merely reflects the fact that the function is defined beyond the range we are interested in. If $p^* > 1$ then 100% of crop x is the least variable combination. If $p^* < 0$, then 100% of crop y is least variable.

10

Planning intercrops – a phenomenological approach

In previous chapters a variety of theoretical approaches, mainly borrowed from the ecological literature, have been presented. For the most part these formulations have been intended as a framework within which intercrops can be viewed, hoping that some qualitative insights of intercrop dynamics might be revealed in the process. In the present chapter and the one that follows we turn to the practical question of how to use such a theoretical framework in the context of intercrop design.

This problem can be approached in two philosophically distinct ways, as has most of the theory already presented. First is what I call the phenomenological approach, common in the ecological literature, in which the problem is formulated as the quantitative response of one species to varying quantities of a second species. Our concern is simply with the quantitative effect of one species on another, specifically ignoring what might be the underlying causes of that effect. Competition and facilitation are thought of as phenomena worthy of study in their own right, irrespective of the underlying mechanisms that produce the observable competitive or facilitative effects.

The second approach is the mechanistic approach, in which competition and/or facilitation are assumed, but the interest is in the study of the mechanisms that cause them. Thus, competition may be for nitrogen or facilitation may be a consequence of protection from herbivores. The mechanism of overyielding might be the partitioning of the nitrogen environment or the mechanism of facilitation might be disrupting oviposition behavior of a key herbivore.

It goes without saying that these two philosophical positions are arbitrary, that a phenomenological approach in one person's eyes is a mechanistic approach for someone else (the 'mechanism' of intercrop advantage is facilitation, or the 'phenomenon' of partitioning the nitrogen environment is caused by some mechanism). For the purposes of this and the next chapter I take the arbitrary position that competition and facilitation are the 'phenomena' which result from underlying 'mechanisms'. Either the phenomena or the mechanisms can be examined profitably, depending on the intended use. The present chapter addresses the phenomena and the following one the mechanisms. In both cases the idea is to elaborate practical ways for engineering intercropping systems.

We begin with an examination of simple competition theory, most of which requires only minor modifications to apply to facilitation. Given that plant competition is at least one of the underlying processes operating in most intercropping systems, it is natural to ask whether it might be possible not only to understand and rationalize the process with ecological theory (as was done in Chapter 3), but also to use that theory to actually plan intercropping systems. Can we use established ecological theory of plant competition to develop a science of engineering intercrops?

That the availability of such technology would be useful is, I think, beyond question. Take, for example, the common problem of determining optimal planting densities. Under normal monocultural conditions the problem is straightforward. Set up a dozen plots with a dozen different planting densities, see which one yields the most, or with a simple equation, interpolate to what must be the optimal density. Now consider the same problem with an intercrop. To get the same level of precision obtained with the monoculture one would have to try all 12 densities of one crop crossed with 12 densities of the second crop, or 144 plots! And that is not even taking into account the large variety of qualitatively different possible patterns (two rows of one crop alternating with one row of the other or single alternating rows or interplanted within rows, etc.). The problem is simply unmanageable if approached with a simple-minded empiricism. Dealing with the question of density in intercropping research has thus always been a major problem (e.g. Francis *et al.*, 1978; Tariah & Wahua, 1985; Fawusi, 1985; Baker, 1978; Bebawi & Abdelaziz, 1983).

Much of the intercropping literature avoids this problem by either dealing with preset patterns (e.g. Vandermeer *et al.*, 1983) or restricting considerations to a question of additive or substitutive patterns (see Chapter 2). While it is common to refer to the so-called substitutive vs additive patterns, this terminology is somewhat unfortunate in that it includes only two of a virtually infinite number of possibilities. Various systematic designs at least partially solve the problem, although they too have their limitations. What we develop in this chapter is, first, an analytical approach to the two-species density problem and, second, a simplified linear approach which does not allow for analytical solutions but might be applicable in a wider variety of situations.

Competition theory for even-aged stands

The following three equations form the basis for the rest of this chapter.

$$w = \frac{k}{1 + aD^{c}}, \qquad (10.1)$$

$$w_i = k - \sum_{j=1}^{m} \alpha_{ij} w_j, \tag{10.2}$$

$$\alpha_{ij} = a\Delta^{-c}. \tag{10.3}$$

Equation (10.1) is one form of the equation that is universally accepted as the best descriptor of the yield–density relationship in plants (Holliday, 1960; Willey & Heath, 1969; Watkinson, 1980; Vandermeer, 1984). Its variables are w, the biomass or yield of an individual plant, and D, the density of the population. It has three parameters k, a, and c, which are usually treated as fitted constants with no obvious biological significance. What we seek is an equation similar to equation (10.1), but including two different densities, to account for the two species in the intercrop. If equation (10.1) has the general form $w = f(D)$, we seek the similar $w = f(D_1, D_2)$. But it is not exactly obvious how this is done. One might simply treat the denominator as a scaling factor and expand it from $1 - aD^c$ to $1 - aD_1{}^c - bD_2{}^c$. Or we could consider the whole right-hand side of equation (10.1) as a single term and simply subtract a similar term to account for the second species. A variety of other options is possible. But the problem is that an examination of equation (10.1) does not provide sufficient understanding of the parameters so as to be able to unequivocally expand them to the two-species situation. The parameters have only been fitted and, with rare exception (Watkinson, 1980), their biological significance is not even discussed. That is where equations (10.2) and (10.3) become relevant.

Equation (10.2) includes the variables w_i, which refer to the biomass or yield of the ith individual in the population, and the parameters, k, the unencumbered yield, α_{ij}, the competition coefficient, and m, the number of individuals in the population. Equation (10.2) simply says that each individual in the population has a maximum potential to yield k. That is, when unencumbered by other individuals, its yield will be equal to k (thus the term 'unencumbered' yield). But each individual in the population is subjected to competitive pressure from the other individuals in the population, of which there are m. So the unencumbered yield k will be reduced by the summation of all the competitive effects from all the individuals in the population.

But implicit in equation (10.2) is the notion that competition is constant from one individual to another. Such an assumption is absurd since two plants growing next to one another will obviously have a greater competitive effect on one another than two plants substantially isolated. That is, the intensity of competition between any two individuals will be a function of the distance between them. This idea is what gives rise to equation (10.3), which relates the competition coefficient α_{ij} to the interplant distance Δ. If the situation is strictly competitive, the function f decreases with increasing Δ. If the situation includes facilitation, the function f likely increases for at least some range of Δ. The exact form of the function f has not been well-verified as of yet, and in fact probably is not strictly general, although the specific form shown in equation (10.3) seems to work on at least some occasions. Intraspecific competition of tomatoes, soybeans, and beans seem to fit it quite well, as does interspecific competition of tomatoes against beans, tomatoes against soybeans, and soybeans against tomatoes (Vandermeer *et al.*, 1984; Vandermeer, 1986). As with equation (10.2), the useful point about the specific form of equation (10.3) is its intuitive sense. It says, quite simply, that the intensity of competition should diminish as a negative geometric function of the distance between competitors, and the parameters thus have very straightforward biological interpretations. The parameter a is the overall intensity of competition (actually the intensity of competition at $\Delta = 1.0$), and c is the rate at which competition declines, or decays, as a function of distance.

Equations (10.2) and (10.3) can be combined, with a bit of mathematical maneuvering (Vandermeer, 1984), so as to result in equation (10.1). That is, the equation that is regarded as the best description of the yield density relationship (equation (10.1)), may be derived from equations (10.2) and (10.3). The attractive part of equations (10.2) and (10.3) is that they have very obvious and intuitive biological meanings.

With these clear biological interpretations it is a simple matter to derive a two-species form of equation (10.1). Extending equation (10.2) to account for an additional species we write

$$w_i = k - \sum_{j=1}^{m} \alpha_{ij} w_j - \sum_{j=1}^{n} \beta_{ij} z_j, \tag{10.4}$$

where β_{ij} is the interspecific competition coefficient, n is the number of individuals of the second species in the population, and z_i is the biomass of the ith individual of the other species. For the second species we have the similar geometric decay of the competitive effect

$$\beta = a\Delta^{-c},$$

where the a and c are different for each species. Putting all these equations together (see Vandermeer *et al.*, 1984 for details), we obtain

$$w_1 = \frac{k - (k_2 a_{12} D_2^{c12} - k_1 a_{22} D_2^{c22})}{1 + a_{11} D_1^{c11} + a_{22} D_2^{c22} + a_{11} a_{22} D_1^{c11} D_2^{c12} - a_{12} a_{21} D_2^{c12} D_1^{c21}}, \tag{10.5}$$

which is exactly the two-species form we were looking for to describe the yield density relationship for two species.

Equation (10.5) has only been used once (Vandermeer *et al.*, 1984) and its performance did not justify its complexity. Despite its cumbersome appearance it is nevertheless tractable and that is the good part. That it is not particularly suited to examining different patterns of plantings (a particular density is considered to be the same whether it occurs as a planting of plants at random or in rows or in clumps), and that it did not perform particularly well the only time it was tried, is the bad part. It is an approach that awaits further empirical conformation. But even if it might be useful in some situations, its complexity and parameter overload suggest that perhaps another approach should be considered.

A computer-based linear methodology

The basic idea

Since equations (10.3) and (10.4) are the initial postulates that eventually give rise to equation (10.5), the question naturally arises, why not use equations (10.3) and (10.4) directly? Actually, equation (10.4) is little more than a statement that competition does occur both within and between species, with the assumption that no higher order effects are significant. Equation (10.4) can be written in matrix form as follows:

$$
\begin{vmatrix} k_1 \\ k_1 \\ \cdot \\ \cdot \\ \cdot \\ k_1 \\ \\ k_2 \\ k_2 \\ \cdot \\ \cdot \\ \cdot \\ k_2 \end{vmatrix}
=
\left|\begin{array}{cccc|cccc}
1 & \alpha_{12} & \cdots & \alpha_{1m} & \beta_{11} & \beta_{12} & \cdots & \beta_{1n} \\
\alpha_{21} & & & \alpha_{2m} & \beta_{21} & & & \beta_{2n} \\
\cdot & & & & \cdot & & & \\
\cdot & & & & \cdot & & & \\
\cdot & & & & \cdot & & & \\
\alpha_{m1} & \alpha_{m2} & & 1 & \beta_{m1} & \beta_{m2} & & \beta_{mn} \\
\hline
\alpha_{m+1,1} & \cdots & & \alpha_{m+1,m} & 1 & \beta_{m+1,2} & & \beta_{m+1,n} \\
\alpha_{m+2,1} & & & \alpha_{m+2,m} & \beta_{m+2,1} & & & \beta_{m+2,n} \\
\cdot & & & & \cdot & & & \\
\cdot & & & & \cdot & & & \\
\cdot & & & & \cdot & & & \\
\alpha_{m+n,1} & \alpha_{m+n,2} & & \alpha_{m+n,m} & \beta_{m+n,1} & \beta_{m+n,2} & & 1
\end{array}\right|
\begin{vmatrix} y_1 \\ y_2 \\ \cdot \\ \cdot \\ \cdot \\ y_m \\ \\ z_1 \\ z_2 \\ \cdot \\ \cdot \\ \cdot \\ z_n \end{vmatrix}
$$

or, in more compact form

$$\mathbf{k} = \mathbf{A}\mathbf{x},$$

where **k** is the vector of k values (the unencumbered yields), **x** is the vector of y and z values (the yields of each individual plant in the intercrop), and **A** is the matrix of interaction coefficients. Note that the matrix **A** is composed of the four submatrices,

$$\mathbf{A} = \begin{array}{c|c} \mathbf{A}_{yy} & \mathbf{A}_{yz} \\ \hline \mathbf{A}_{zy} & \mathbf{A}_{zz} \end{array}$$

where $\mathbf{A}_{ki}$ is the matrix of the interaction coefficients which stipulate the effect of individuals of species i on species k.

If the *k*s and the interaction coefficients are known, we can predict the yields with the equation

$$\mathbf{A}^{-1}\mathbf{k} = \mathbf{x}, \tag{10.6}$$

which is derived by multiplying the matrix equation by the inverse of **A**. Equation (10.6) is the theoretical basis of the predictive methodology described here. With a small series of experiments it should be possible to estimate all the parameters of **k** and **A**, for a particular crop combination under specified conditions, and then use equation (10.6) to predict the yields of both crops (i.e. the elements of the vector **x**) under a variety of possible planting designs.

But changing the planting design has an obvious and important effect on the interaction coefficients themselves. A neighbor plant with its roots intertwined with a first plant will have a bigger effect on that plant than some neighbor several meters away. For this reason equation (10.3) is necessary, in its specific geometrical form, or as some other specific form of the function *f*. Equations (10.3) and (10.6) stipulate all that is necessary to predict intercrop yields. For each species it is necessary to estimate empirically the three parameters *k*, *a*, and *b*. The proposed physical design of the intercropping system then fixes the values of delta. From the proposed values of Δ and the known values of *a* and *b*, the interaction matrix is calculated. Premultiplying the **k** vector by the inverse of the interaction matrix (equation (10.6)) provides the predicted yields of the proposed system.

The basic algorithm, then, can be represented in step form as follows:

1 enter coordinates for each individual plant (of both species), for the design in question;
2 compute the deltas;
3 compute the αs and βs (equation (10.3));
4 invert the interaction matrix;
5 compute the predicted yields (equation (10.6)).

Using this methodology, a researcher is able to screen large numbers of potential designs very quickly. While it is not possible to actually compute optimal designs or densities, it is easy to predict yields for a large number of possible designs. Thus, a researcher or planner can sit in front of a monitor, put a particular design up on the screen and quickly view the predicted yields, then change the design and predict the yields again, change the design again, and so on.

Stabilizing the yields

Solving equation (10.6) is simple with a small population. With a larger population one must use a computer. But even with a modern computer, when

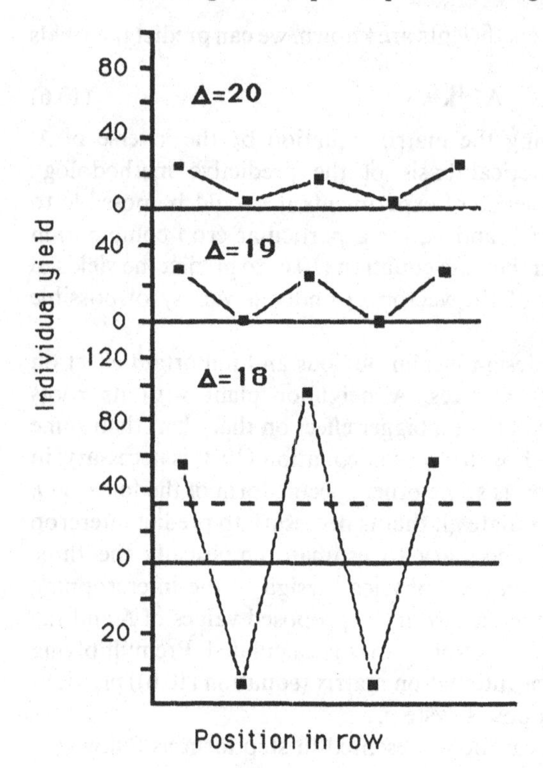

Fig. 10.1. Individual tomato yields as a function of position in the row, for interplant distances of 20, 19, and 18 cm. Dashed line is theoretical maximum yield (from Vandermeer, 1986a).

the population contains more than 200-300 individuals, practical problems of the cost of computer time and/or the introduction of round-off error (in inverting the matrix) arise. The intention of the method is thus not to simulate a full hectare or some other large unit, but rather to compute yields for a smaller sample of the field, as small as possible such that the basic pattern of the overall planting design is preserved in the sample. But a small unit brings up the question of edge-effect.

Even in a monoculture where the parameters of the model are identical for all individuals, not all individuals perform identically since they are in different physical positions and therefore different competitive relations with one another. Figure 10.1 shows this effect dramatically for predicted tomato yields. The plants on the edge of the row, receiving competitive effects from only one side, have a much greater yield than any of the others. This creates a situation in which the second plant from the edge does very poorly since one of its major

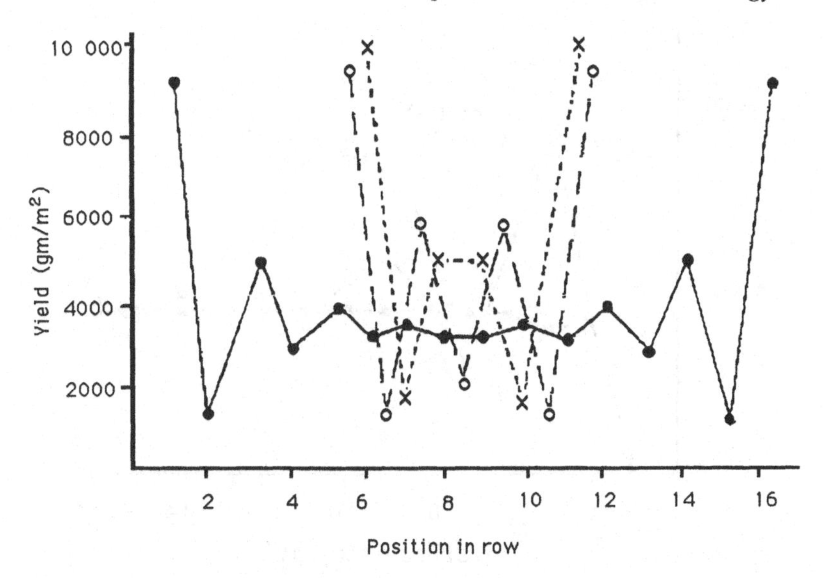

Fig. 10.2. Stabilizing the yields. With six (×) or seven (o) individuals in the row the yields are dramatically influenced by artificial edge-effects. With 16 individuals (solid dots) in the row, the central individuals all yield approximately the same (data from tomato yields, Vandermeer, 1986a).

competitors is the edge plant which is quite vigorous. The pattern repeats itself as we move from the edge to the center of the row, but with decreasing amplitude. The whole process is illustrated in Figure 10.2. When there are only six or seven individuals in a row (dotted and dashed lines in Figure 10.2), the variation at the edge is just as severe as with the smaller population, but towards the center of the row it is substantially reduced. Obviously we seek to estimate the yields of the plants in the center of the row, not the highly variable individuals on the edges.

A solution to the edge-effect problem might simply be to simulate a sufficiently extensive area so that edge effects will be minimized, a procedure that works fine in a monoculture. But in an intercrop we are frequently forced to deal with two crops of very different normal planting density. This results in unreasonable numbers of the more dense species being required for simulation. For example, in predicting the yields of a tomato–bean intercrop (as described below) one of the planting patterns is rows spaced 1 m apart, with tomatoes spaced 1.0 m within rows and beans 2.5 cm within rows. If we insist on 16 (say) tomatoes per row, that means we require 600 beans per row, and a minimum of 3 rows requires $616 \times 3 = 1848$ total plants. Inverting an 1848×1848 matrix is somewhat cumbersome.

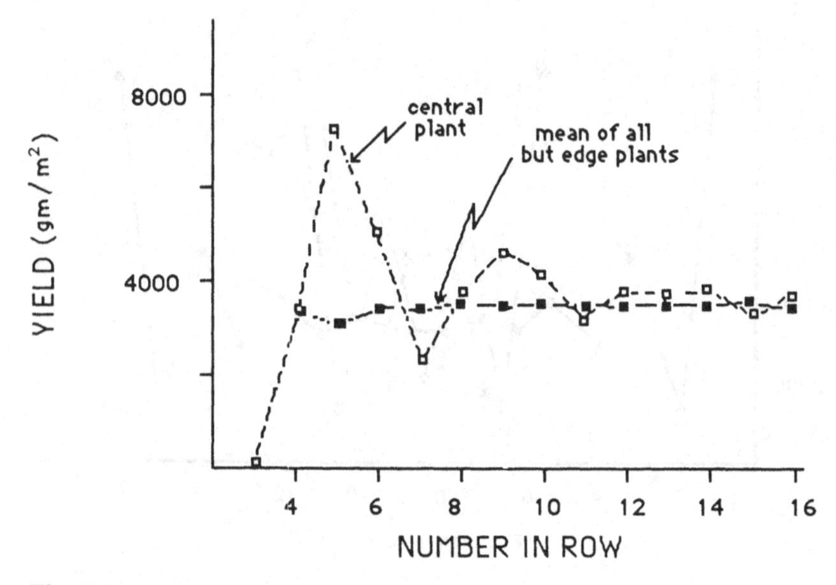

Fig. 10.3. Yields of all plants in the row, as a function of number in row. Open squares are individual yields of the central plant, closed squares are the mean values of all but the extreme edge plants (from Vandermeer 1986a).

Fortunately, there appears to be a convenient way to estimate the 'stabilized' yield short of including a large number of individuals in the population. Figure 10.3 shows the per plant yields of simulated populations of tomatoes for various numbers of individuals in the row. In this diagram the yield of the central (or two central) plant is indicated by the dotted connecting line. As can be seen, the yield of the central plant stabilizes at about 3400 gm. Also plotted is the mean value of all but the edge plants. At about five or six plants in the row this value has apparently already stabilized between 3200 and 3600. It would appear that one can obtain a good approximation of the stabilized value by computing the mean of all but the edge value for a row of five or six individuals. That is the procedure recommended in general until such a time as its failure is demonstrated.

Estimating parameters

The intraspecific parameters may be estimated in two ways, indirectly and directly. The indirect technique involves the standard density trial experiments in which one obtains per unit area yields at a range of different densities. These yields can be fit to a modified form of the yield–density equation (Watkinson, 1980; Vandermeer, 1984)

$$Y = \frac{Dk}{1 + aD^{c}},$$

where k is the theoretical maximum yield attainable by an individual plant, and the parameters a and c are essentially equivalent to a and $-c$, respectively of equation (10.3) (Vandermeer, 1984). Thus, all parameters of the model are given by a fit to the modified yield–density equation. Since it is not possible to linearize the equation, some nonlinear technique must be used to fit the data to the model.

The second method, here called the direct technique, is to estimate all three parameters (k, a, and c) from target experiments (Goldberg & Werner, 1983; Goodall, 1960; Mead, 1979; Vandermeer *et al.*, 1984), which are modifications of the older more ambitious technique of beehive experiments (Veevers & Boffy, 1975; Martin, 1973). An individual plant, the 'target' is surrounded by a number of other plants, the 'edge plants' or 'effector plants'. The effector plants are arranged around the target in such a way that each individual effector plant is the same distance from the center plant. The distance from the center plant to the edge plant is called the 'interplant distance'. It is an important assumption here that the center plant receives the effects (competition or facilitation) from only the effector plants, not from any other plants in the vicinity (Vandermeer, 1984). Thus, not only the area inside, but also the area outside the surrounding plants must be kept clear of living vegetation. Figure 10.4 illustrates target experiments with maize and beans.

A simple modification of equation 10.2 yields

$$\alpha = (k - y_c)/my_e, \tag{10.7}$$

where y_c is the yield of the target, y_e is the yield of an average effector plant, and m is the number of effector plants. Note that k must be estimated independently, easily accomplished with several isolated plants.

Equation (10.7) is applied at each interplant distance, resulting in a series of values, one for each interplant distance. Rewriting equation (10.3) as

$$\ln\alpha_{ij} = \ln a - b\ln\Delta_{ij},$$

we have an equation that can be used with the experimentally obtained values to estimate a and b through linear regression. If the geometric form is not applicable (i.e. if the data do not fit equation (10.3)), some other fitting technique must be used to estimate whatever parameters appear in the function f.

An example

This technique was first applied to a system of tomatoes and beans in southern Michigan, U.S.A. (Vandermeer, 1986). Four sets of target experiments were set

Fig. 10.4. A 'target' experiment in which a maize plant is the target and is surrounded by 'effector' bean individuals.

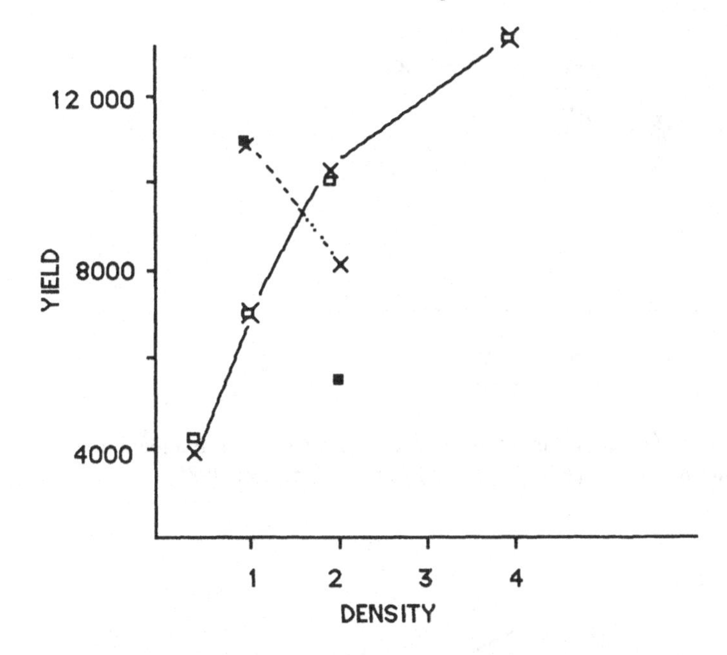

Fig. 10.5. Predicted (× s) and observed (squares) yields of tomatoes in monocultures (solid lines) and intercrops (dotted line) (data from Vandermeer, 1986).

up to estimate all the parameters of the model. Then, a series of plot trials were set up so as to check on the model's ability to predict yields in a row cropping situation.

The general performance of the model is presented in Figures 10.5 and 10.6. Figure 10.5 shows the predictions that were made for the tomatoes, and Figure 10.6 for the beans. The model performed quite well in this case, at least as judged by the figures. Individual analyses of variance showed a range of 59–100% of the variance explained by the model (see Vandermeer, 1986, for details).

But these data, both those cited above and those presented in Figures 10.5 and 10.6, leave out an important part of the problem. Many of the intercropping combinations gave no prediction at all with the model. Frequently, a predicted set of yields included negative values, a consequence of strong competition in a linear model. Such a prediction clearly is biologically ridiculous, but stems from a real biological process. That process is called self-thinning. Long recognized as an important factor in plant ecology (Yoda *et al.*, 1963; White, 1981; White & Harper, 1970; Westoby, 1984), self-thinning becomes especially important in the matrix projection. The generalities of

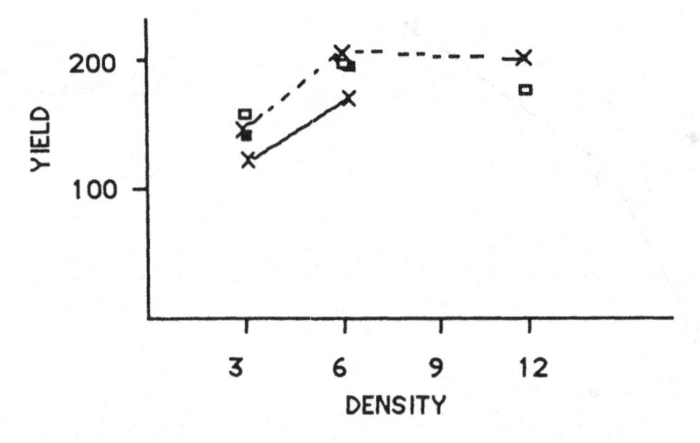

Fig. 10.6. Predicted (x) and observed (squares) yields of beans in monocultures (dotted line) and intercrops (solid line) (data from Vandermeer, 1986).

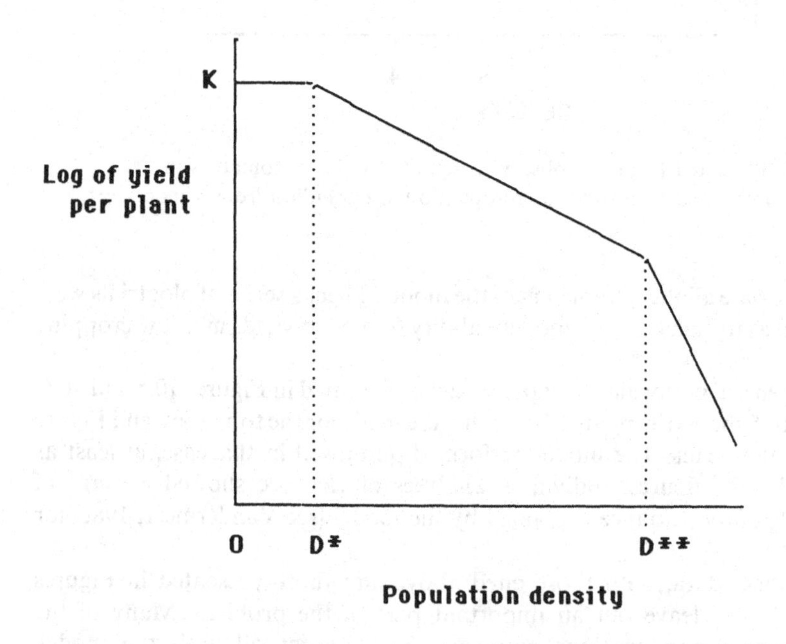

Fig. 10.7. General relationship between the log of yield per plant and population density (see text for explanation).

yield and mortality relationships with regard to density are pictured in Figure 10.7. The process can be thought of as three linear segments. First, densities are so low that each plant in the population grows unencumbered (between 0 and D^* in Figure 10.7). Second, yield is reduced by competition but there is no mortality (between D^* and D^{**}). Finally, competition is so intense that mortality (thinning) occurs (above D^{**}). While the actual relationship is likely to be a smooth monotonic function (e.g. Slatkin & Anderson, 1984), it is convenient to draw it in three linear segments so as to emphasize the three separate processes involved.

The simple linear model described here probably applies to situations generally situated between D^* and D^{**} of Figure 10.7, that is, when thinning (mortality) is not an important consequence of either intra or interspecific competition. The cases that failed to give predictions in the above example were all examples of planting patterns that were sufficiently dense to produce mortality (i.e. at densities greater than D^{**} of Figure 10.7). Because of the way in which the model is constructed, a thinned individual appears to have a negative biomass, absurd biologically, but even worse, that individual with negative biomass contributes positively (a negative biomass times a negative competitive effect is a net positive effect) to its neighbors.

The solution to this problem is perhaps simply to avoid it. It might seem that in any practical situation it will rarely be advantageous to sow above densities that generate self-thinning. Restricting the application of this model in this way, it is only necessary to compute the critical self-thinning interplant distance and restrict the model's usage to only larger distances. On the other hand, if higher densities are desired for some reason, the present method simply will not work, and an alternative technique must be sought. This problem is discussed in Chapter 12.

11

Planning intercrops: a mechanistic approach

Introduction

While the phenomenological approach of the previous chapter was conducive to generalization, it is, almost by definition, impossible to generalize about the subject of this chapter. Since the mechanistic view includes such a wide variety of topics, from insect movements to shading to resource utilization, no single methodology is possible. Nevertheless, it is possible to group mechanisms into broad categories and discuss potential approaches within each of these categories.

Following the overall structure of this book, the mechanistic approach is presented in two broad categories, mechanisms of reduced competition and mechanisms of facilitation. Within the category of reduced competition are included two topics: partitioning the light environment, and partitioning resources. For the same reason that it was possible to develop an analytical approach to the environmental impact of one species on the environment of the other, in chapter 7 a potential engineering approach is possible when competition between two annuals is known to be for light. Since light is unidirectional, there are certain inevitable rules upon which one can build, at least provisionally. On the contrary, when competition is for resources no such inevitable rules are obvious.

Within the category of facilitation, above and beyond the clear planning implications of Chapters 4 and 6, which are not repeated here, there are three topics elaborated, all associated with pest control. First, the so-called 'trivial' movements of pests and predators, which is to say their movement patterns within a field after having arrived, are treated with the standard application of diffusion equations. Second, the question of entering and leaving a crop, immigration and emigration, is examined with an eye to the effect of trap-crops on those processes. Third, if the potential for weed control exists (see Chapter 8), the question of designing the system to optimize this control arises.

The following material is presented in the spirit of 'tactical' modeling (see Chapter 1), as was the phenomenological approach of the previous chapter. However, the material in this chapter has not seen the same sort of empirical justification. The mechanistic approach is necessarily more eclectic, and relies on more knowledge about the intercrop of concern than does the phenomenological approach. Despite the lack of empirical justification it seems worthwhile to present these few examples of what tactical models based on underlying ecological principles might look like. Their presentation is thus meant to be in a spirit of tentative suggestions rather than fully verified technologies.

Mechanisms of reduced competition

Partitioning the light environment

It is often presumed that producers choose crops based on the differential utilization of the light column. In those cases in which this mechanism operates, it is feasible to design the intercropping system based on some elementary principles. As already introduced in Chapter 5, the idea is to replace the lower leaves of the taller crop, presumed to be a 'sun' species, with leaves of a shorter crop, presumed to be a 'shade' species. If the lower leaves of the tall species were operating below compensation, as frequently is the case, the system as a whole will be more productive with a short-shade species taking the place of the nonproductive lower leaves of the tall species (see Figure 5.2).

The system can be visualized as an effect–response combination (see Figure 1.1). A plant, or plant population, responds to a light environment that has been affected by itself and the other species. The problem is well-known to the crop physiologist when dealing with monocultures. A very dense stand has a large number of plants but creates a light environment in which production of lower leaves is low. A very sparse stand creates a light environment that is conducive to high production of all leaves, but fewer plants to do the producing. The very same idea functions in an intercropping situation but involves a more complex set of factors, since two species are involved.

Each species has an effect on the environment which is to reduce the amount of radiation reaching all but the top leaf layers. There are actually three variables involved, y = the position in the light column (height above ground), x = the foliage density (one measure of which is the leaf area index), and z = the radiation intensity. The foliage density is a function of y (here we consider the foliage density as a cumulative index increasing from a value of zero above the crop), and the radiation intensity is a function of the foliage density. For a monoculture we would have $x = f(y)$ and $z = g(x)$. The 'effect' can thus be represented as the composite function $h = g[f(y)]$. These rules apply equally to intercrops, but with certain complications.

When intercropping the variables are conceptually the same, but are combined in slightly different ways. Thus we have the two foliage densities (one for each crop) as they appear in the intercropping situation, as $x_1 = f_1(y)$ and $x_2 = f_2(y)$, representing the foliage density from the top of the canopy to the level y, and $z = g(x_1 + x_2)$ = the radiation intensity dictated by the foliage density x, specified for each of the two species. The radiation intensity is also a function of the position in the canopy ($z = g(x_1 + x_2) = h(y)$), as it was formulated in Chapter 7. But here we are interested in how interspecific competition modifies the foliage density, $x_i = f_i(y)$, so we analyze the phenomena as two separate processes, $x_i = f_i(y)$ and $z = g(x_1 + x_2)$.

It is worth emphasizing that the whole point of formulating the problem in this way is to be able to stipulate how the functions f_1 and f_2 change as a function of planting pattern. A very dense pattern will show a quantitatively different form of f than will a sparse one, a high relative abundance of one species a different form than a low relative abundance, etc.

Finally, a particular radiation intensity dictates photosynthesis for each species. Thus, p = photosynthesis, $p_1 = m_1(z)$, and $p_2 = m_2(z)$ for species 1 and 2 respectively. These latter relationships can then be formulated in terms of the position in the canopy (e.g. $p_1 = m_1(h(y))$), which casts the whole problem in terms very similar to the ideas of niche theory, as outlined in Chapter 3. A measure of 'fitness' is P and the resource gradient is the vertical position in the canopy. The fundamental niche is thus $p_i = m_i(f_i(y))$ and the realized niche is $p_i = m_i(h(y))$.

With this formulation we see that

$$LER = \frac{m_1(h(y))}{m_i(f_i(y))} + \frac{m_2(h(y))}{m_2(f_2(y))}.$$

The practical engineering problem then is to maximize LER, an exercise whose difficulty will depend in large part on the exact forms of the functions f, g, and m. I know of no example in the literature in which data are available to estimate all three functions, but the simple logic of the situation suggests this program as an interesting avenue for future research.

On the other hand, I do not wish to imply that the difficulties of using this method are not severe. First, establishing the relationship between foliage density and height must be done for each species of concern, in monoculture at several densities and then in the intercropping situation at various density combinations. It is possible to presume random distributions of leaves and utilize a Poisson distribution to derive a general formula. But while such an approach might be conducive for developing generalized qualitative models (e.g. Horn, 1971), I doubt that it would suffice for the sort of engineering applications envisoned here.

Second, the function g is also not easily stipulable. While it is common to apply Beer's law to this function (e.g. Saeki, 1960; Monsi & Saeki, 1953),

Monteith (1965) proposed a more complicated function and Stanhill's data on alfalfa (Stanhill, 1962) were well represented by a simple linear function. As with the function f, different forms of g will probably have to be applied to different crop combinations.

Third, the function m will probably be less of a problem since the Michalis–Menton form appears to describe most available data quite well (e.g. Horn, 1971; Monteith, 1965).

Finally, the time course of this process will likely have to be taken into account, a problem ignored in this initial outline.

Nevertheless, an engineering application based on the simple proposition that we can analytically describe (1) the changes in foliage density patterns as a result of competition (the function f), which leads to (2) a description of the way shade is produced (the function g), which enables us to (3) predict the photosynthetic response of each species, is an intriguing outline, worthy of further research.

Partitioning of the resource environment

Competition for resources is conceptually similar to competition for light in that the resource pool is modified by each of the crop species, which in turn respond to the modified resource pool. What is fundamentally different is that there is no obvious anatomical analog to foliage density that might be used to develop the corresponding theory. While it might be tempting to use total root density as a function of soil depth, the difficulty of measuring such a variable, together with the questionable (at best) relationship between root density and resource uptake, mitigates against such an approach. Short of developing a dynamic methodology, as further discussed in Chapter 12, I know of no extant ecological theory, nor can I conceive of one that might be modified to apply to the question at hand. The work of Hsu *et al.* (1977) and Tilman (1982), while explicitly concerned with species coexistence, nevertheless are suggestive of more sophisticated developments.

Mechanisms of facilitation

The general approach taken in Chapter 4 is relevant to practical applications involving the planning of intercrops. If a principal crop is subjected to both competitive and facilitative effects by a secondary crop, and those effects depend on the level or intensity of that secondary crop, then it is simply a question of balancing the competitive effect and the facilitative effect so as to maximize the production of the principal crop. The technology for doing this has already been described in detail in Chapters 4 and 6. Here the concern is with modeling the specific modifiers that create facilitation in the first place.

We deal with three specific topics, the use of diffusion equations to describe pest and natural enemy dynamics, the design of trap-crops, and the use of intercrops to alleviate weed problems.

Pest and natural enemy dynamics

As described in detail in Chapter 6, it is well known that intercrops frequently demonstrate a lower level of pest attack than monocultures. It should be an important goal of intercropping research to develop means of informed manipulation to improve on this aspect of intercropping advantage, thus making intercropping a potentially important aspect of integrated pest management programs (Speight, 1983; Vandermeer & Andow, 1986; Rosset, 1986). It is generally thought that one of three mechanisms are responsible for observed patterns: the disruptive crop effect, the enemies effect, or the trap-crop effect (see Chapter 6). These mechanisms ultimately involve movements of the pests in question, and thus are amenable to analysis with diffusion equations, as argued by Kareiva (1986). We thus begin with a description of the application of diffusion equations to the problem of intercropping.

The basic idea can be seen intuitively by considering a population dispersed along a one-dimensional gradient, as, for example, along a row of some crop. If we suppose there is no reproduction or mortality, the change in the number of individuals at any point along the gradient will be a function of only those individuals diffusing onto that point and those diffusing away from that point. So the rate of change of the population at a point will be proportional to the rate at which individuals reach the point minus the rate at which individuals leave the point. The rate of arrival minus the rate of departure is simply the rate of change of the rates, or the second derivative. Thus we have

$$\frac{\partial N}{\partial t} = D\frac{\partial^2 N}{\partial t^2}.$$

This is the elementary diffusion process, the solution to which is a normal distribution.

But the intent here is to apply the same concept in a two-dimensional crop field. The basic ideas are the same, but the equation gets a little more complicated. Specifically, if we wish to examine the rate of change of a herbivore population in a two-dimensional situation, we may write

$$\frac{\partial H}{\partial t} = D_x\frac{\partial^2 H}{\partial x^2} + D_y\frac{\partial^2 H}{\partial y^2} - eH, \tag{11.1}$$

where H represents the local density of the herbivores, the Ds represent diffusion coefficients in the two coordinate directions, and e represents the disappearance rate.

Table 11.1. *Rates of leafhopper movement and disappearance*

Treatment	Movement rate		Disappearance rate
	Along rows	Across rows	
Low-density maize monoculture	0.031	0.054	0.129
High-density maize monoculture	0.027	0.494	0.125
Maize–bean intercrop	0.062	0.036	0.292

(Data from Power, 1987.)

Dempster (1957) provided a general method for estimating the partial derivatives from simple distribution data, which made it possible to compute the diffusion coefficients (movement coefficients) and the disappearance rate. Power (1987) used this method to compare movement rates of the leafhopper *Dalbulus maidis* in maize monocultures and maize intercropped with beans, some of the results of which are presented in table 11.1. The rate of movement along rows and the disappearance rate were twice as rapid in the intercrop as in the monoculture, but the rate of movement across rows was dramatically reduced in the intercrop. As would be expected from the disruptive crop, movement along the rows was more rapid when an effective block of beans was between the rows, suggesting a more rapid rate of disappearance in the intercrop.

The application of this technique for planning intercrops specifically for protection against a mobile pest is obvious. The two diffusion constants and the disappearance rate can be measured experimentally at different planting patterns so as to develop an empirical relationship between the diffusion coefficient and planting pattern. Solving equation (11.1) for various diffusion values could then provide the optimal pattern for either maximizing the disappearance rate or minimizing the rate of diffusion of the pest through the field.

While this simple diffusion approach may be useful for many pests, there are other situations in which the time of attack of the pest, with respect to the growth of the crop, will be of critical importance. For a monoculture, Kareiva (1986) suggests the use of simultaneous differential equations, with diffusion added. For example, the growth of the crop might be

$$\frac{\mathrm{d}P}{\mathrm{d}t} = \mu P(1-P) - (aHP)^k,$$

and the attack of the pest

$$\frac{dH}{dt}=rH[1-H/(cP)],$$

to account for local dynamics and the movement of pests (μ and r are growth rates, k and c are constants, and H and P refer to the population densities (or biomasses) of the herbivore and predator, respectively).

This approach, when applied to an intercropping situation, becomes much more complex in that the equation describing the crop growth must be replaced by two simultaneous equations, something yet to be successfully accomplished in the theory of plant competition. This problem is laid out in considerable detail in Chapter 12.

The problem becomes even more severe if natural enemies (predators) are involved. The predators must be modeled as a diffusion process also, but the diffusion constants must be considered not just functions of the positions of the plants but also of the local population densities of the prey (pests). While much progress has been made in developing the diffusion approach to predator–prey interactions (e.g. Banks *et al.*, 1987), coupling such an approach to interacting plant populations has not been tackled so far.

Trap-crops

Thus far in this section the concern has been with only the trivial movements of pest organisms, that is, the movement patterns within an intercrop and monoculture, after the larger events of immigration and emigration have been factored out. In many cases the more important issue will involve those larger problems of immigration and emigration. In the case of trap-crops this is true. As to whether a given situation warrants trap-cropping in the first place has already been covered in Chapter 6. Here we consider the simple engineering problem of designing a system in which the efficiency of the trap is optimized.

Consider first the possible competitive effect against the crop from the trap strips. The ith crop is expected to yield k when no competitive effect is operative, but when the full effect of uncontrolled herbivores is felt, k will be reduced by some amount with each affecting strip. We suppose the general linear form

$$y=k-\alpha(\Delta),$$

where α is the competition coefficient and is a function of Δ, the distance between the affected row and the trap.

Consider the yield of the ith crop row. In general we can write

$$y_i=k-\sum_{j=1}^{\infty}\alpha_j(\Delta) \tag{11.2}$$

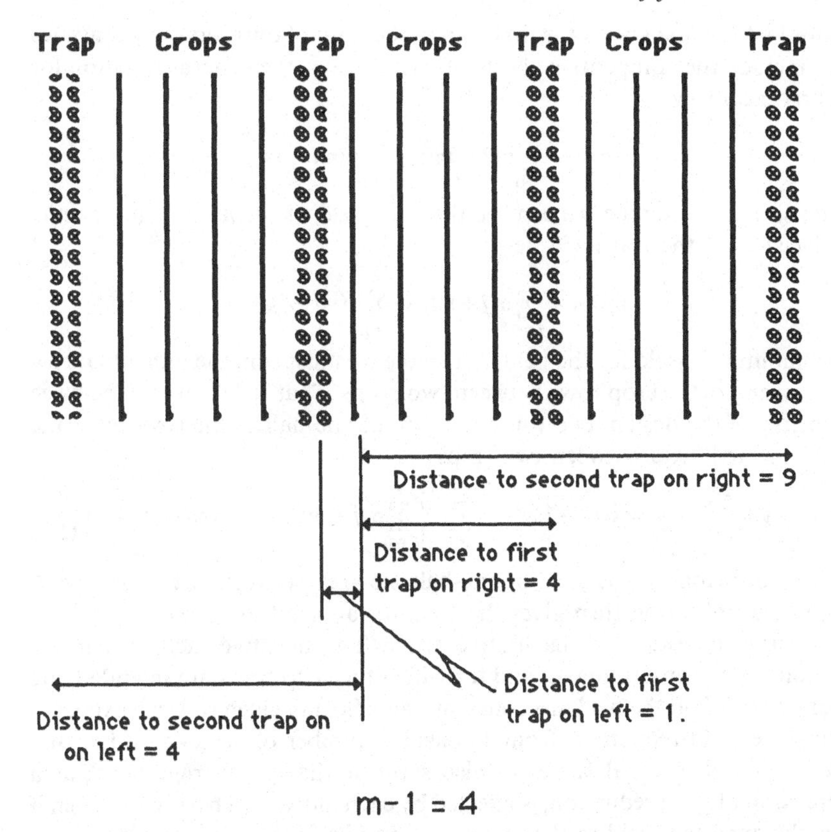

Fig. 11.1. Diagrammatic representation of the distances from trap to crop row, in a trap-cropping system.

where the summation is taken over all the traps. Naturally there will be some real limit beyond which the competitive effect is zero. The question here is how to compute the summation term in equation (11.2) given that the relationship between α and Δ is known either analytically or as a tabled function. Let $m-1$ be the number of crop rows between two traps. The first crop row to the right of a trap will be 1 unit distance from that trap (on the left of the crop row) and $m-1$ units away from the trap on the right and $m-1+m=2m-1$ units distance from the next trap further on the right. These distances are illustrated in Figure 11.1.

Thus, we may in general write, $(1+im)$ for the distance to the ith trap removed from the nearest trap on its other side (e.g. in Figure 11.1 the first trap is actually the second one to the left), and $(im-1)$ for the distance to the ith trap removed from the nearest trap, on the same side as the crop row (e.g. in

Figure 11.1 the first trap is the first trap to the right of the crop row). Since we must include the competitive effects of all the traps, the general equation for the first crop row is

$$y_1 = k - \sum_{i=0}^{\infty} \alpha(1+mi) - \sum_{i=1}^{\infty} \alpha(mi-1).$$

Using similar reasoning we can see that the yield of the jth row (i.e. j units displaced from the trap) will be

$$y_j = k - \sum_{i=0}^{\infty} \alpha(j+mi) - \sum_{i=1}^{\infty} \alpha(mi-j).$$

To compute the yield for the field as a whole we must sum the individual row yields over $\frac{1}{2}$ of the crop rows between two crops. That is, because of the basic symmetry of the design, two times the sum of one-half of the rows gives the total expected yield between two traps

$$y = \sum_{j=1}^{\frac{1}{2}(m-1)} y_j = \tfrac{1}{2}(m-1)k - \sum_{j=1}^{\frac{1}{2}(m-1)} \left\{ \sum_{i=0}^{\infty} \alpha(j+mi) + \sum_{i=1}^{\infty} \alpha(mi-j) \right\}, \quad (11.3)$$

as the expression for the yield of one-half the area between any two traps. A simple multiplication then gives the yield for an arbitrary area.

We now consider the facilitative effect. The negative factors that are normally in the environment, and for which the strip traps are intended, are assumed to fall on the field in a constant 'rain'. So any given row can expect to experience yield reductions from a constant number of herbivores. Let that constant number equal h. Let us also suppose that an average of 'a', as a proportion of yield reduction, is affected by each individual herbivore. Then if k_m is the expected yield in the absence of any herbivore, we can write

$$k = k_m ah$$

as the expression of the unencumbered yield – the yield expected when competition from the traps is nonexistent.

We now must examine how h can be reduced through its proximity to a trap. While h is the actual number of herbivores in a row, we define h_m as the maximum number that would be in the row if no disruptive effect were felt from the traps. The actual number in the row will be some fraction of that maximum, or

$$h = \beta(\Delta)h_m,$$

where $\beta(\Delta)$ is the fraction and is construed as a function of the distance to the trap. Considering the crop row immediately adjacent to the trap, we can write

$$k = k_m \beta(1) h_m a$$

to account for the proportional reduction from just the adjacent traps. When the other traps are also added, fractionally reducing the number, using the same spatial logic as before, we obtain

$$k = k_m h_m a \prod_{i=0}^{\infty} \beta(1+mi) \prod_{i=1}^{\infty} \beta(mi-1)$$

for that row. Generalizing to the first half of the rows between two traps we obtain

$$y = k_m h_m a \sum_{j=0}^{\frac{1}{2}(m-1)} \prod_{i=0}^{\infty} \beta(j+mi) \prod_{i=1}^{\infty} \beta(mi-j). \tag{11.4}$$

Combining equations (11.3) and (11.4) we obtain the general equation

$$y = \tfrac{1}{2}(m-1) k_m h_m a \sum_{j=0}^{\frac{1}{2}(m-1)} \prod_{i=0}^{\infty} \beta(j+mi) \prod_{i=1}^{\infty} \beta(mi-j) - \sum_{j=1}^{\frac{1}{2}(m-1)} \left\{ \sum_{i=0}^{\infty} \alpha(j+mi) + \sum_{i=1}^{\infty} \alpha(mi-j) \right\},$$

which can be used directly to compute the yield between rows for any trap design. Maximizing y with respect to m gives us the optimal trap design.

Planning for weed control

A final consideration is the indirect facilitation of one crop by a second through the second crop's effect on the weed community. In the context of using a cover crop, or a nonvaluable species specifically to control weeds in a principal crop, we now examine some design problems centering on the problem of the critical planting time and the critical weed-free period (Unamma *et al.*, 1986; Lolas, 1986). The model for this problem is seen in many areas of West Africa in which a single weeding is performed about a month after planting sorghum, millet, or maize. Immediately after weeding, a legume crop is planted between the rows, thus combining the activities of weeding and seed-bed preparation (Steiner, 1984). The engineering problem is to determine the optimal time to weed and subsequently plant the secondary crop.

We begin the analysis by assuming the goal is to control the weed, and not to receive any monetary benefit from the crop that is doing the controlling. This simple case is likely to be the most commonly encountered, and actually corresponds to the normal practice of using a cover crop, although the present analysis is restricted to an annual cover crop (herein called a control crop). We further presume that the control crop is known to reduce the weed to a level that will not affect the crop, if the weed and control crop initiate their growth simultaneously. The variable we wish to analyze is the time at which the control plant should be planted, stipulating that the system is maintained in a weed-free condition prior to that planting time. Thus the system includes a harvestable crop, a control crop, and a weed, and the object is to stipulate the optimal time to plant the control crop.

Suppose that biomass accumulation in the control crop is given as the simple exponential function, $dB/dt = rB$, where B is the biomass of the control crop, and that weeds are controlled by some external force up to the point at which the control crop is introduced (i.e. there is a weed-free period at the beginning of the cropping cycle). Suppose there is a parameter C, which stipulates the critical biomass, above which the control crop itself has an unacceptable effect on the crop. Since the crop itself is growing, the level the control must attain to affect the crop will increase with time. Thus C is a monotonically increasing function of DAP (days after planting).

Similarly, the control crop will grow to some level, its 'attainable biomass'. That biomass will be at least partly conditioned by the biomass of the principal crop, which itself competes with the control crop. Thus K, the attainable biomass, is also a function of DAP, but a decreasing function thereof. The system as a whole is stipulated by the following three equations:

$$dB/dt = rB,$$
$$C = f_1(DAP),$$
$$K = f_2(DAP),$$

where f_1 is monotone increasing and f_2 is monotone decreasing. Furthermore, we presume that there exists a value of DAP below which $C < 0$.

The first problem is to stipulate the time (DAP) at which the actually attained biomass (K) is less than or equal to C. Set $K = C$ as the critical point. That is, we seek a value of DAP (DAP^*) such that

$$f_1(DAP^*) = f_2(DAP^*),$$

and considering the dynamics of the growth of the control crop,

$$\ln B = a + rt,$$

where a is the log of the biomass when $t = 0$. Substituting the value of the biomass at the critical point, we obtain

$$\ln[f_1(DAP^*)] = a + rt^*,$$

where t^* is the critical planting time (i.e. the time at which the control crop must be planted so as to reach critical biomass at exactly DAP^*). Rearranging, we write

$$\frac{\ln f_1(DAP^*) - a}{r} = t^*.$$

In Figure 11.2 these definitions are pictured diagrammatically.

In practical terms these ideas must be translated into economic terms. There are two costs, the cost associated with weed control and the cost associated with crop loss due to competition from the secondary or control

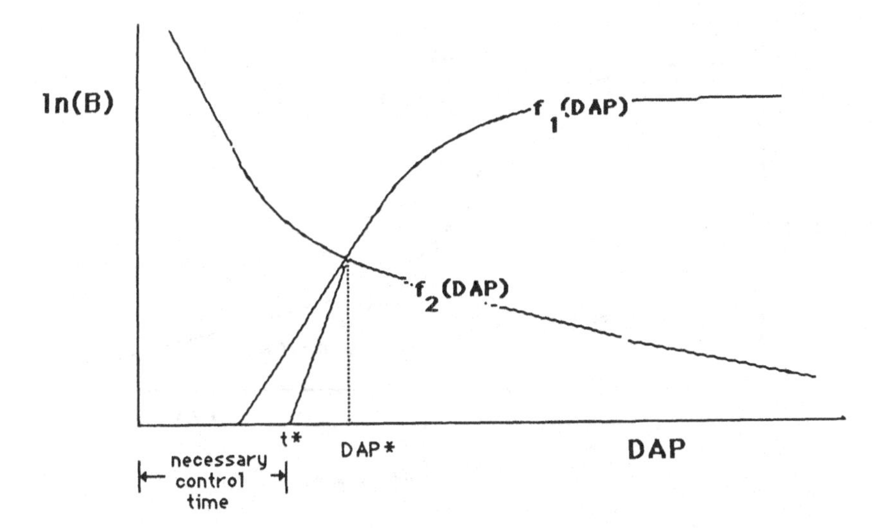

Fig. 11.2. Relationship between log biomass and time, showing the position of the necessary weed-control time.

species. The cost associated with the control procedure can be visualized as a cost per unit time. If the time necessary to maintain a weed-free crop is large, the cost of weed control is large. Thus the cost of weed control to the point of planting the secondary species will be nt, where n is the cost per unit time.

The cost due to losses exerted by the control species will be proportional to the difference between the attained biomass of the control crop (K), minus its critical biomass (C), or $m(K-C)$, where m is the cost exerted by a unit biomass of the control species above its critical biomass. The total cost, T_c, is simply the sum of these two components, or

$$T_c = nt + m(f_2 - f_1),$$

whence

$$dt_c/dt = n + m(df_2/dt - df_1/dt).$$

Total cost is minimized when $d(f_2-f_1)/dt = 0$ and $d^2(f_2-f_1)/dt^2 > 0$, thus

$$d(f_2 - f_1)/dt = n/m,$$
$$d^2f_2/dt^2 > d^2f_1/dt^2,$$

or, when the difference in the derivatives of the two functions is equal to the ratio of cost per unit time and cost per unit biomass. This condition is pictured in Figure 11.3, along with the extrapolations of what would happen if prices changed. If the cost of weed control increases, the ratio n/m increases and the optimum planting time decreases. If the value of the crop itself increases, and

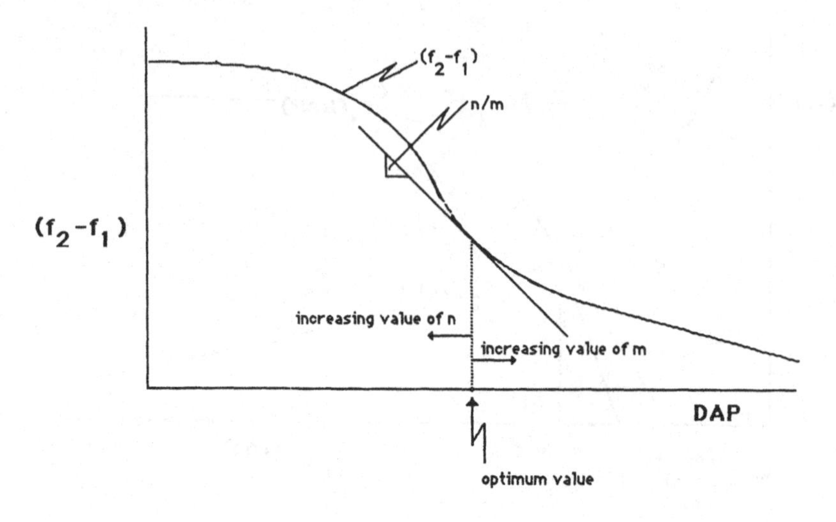

Fig. 11.3. Optimizing weed control and expected changes in optimum as prices change (see text).

thus the cost of the attained biomass of the secondary crop increases, n/m decreases and the optimum planting time increases.

Summary and conclusions

Using a mechanistic approach to the design of intercrop systems presumes that a great deal is known about how they function in the first place. Unlike the phenomenological approach of the previous chapter, here the concern has been with using the understanding of specific mechanisms to engineer intercrop improvements. Since it is so infrequently the case that a full understanding of intercrop advantage is available, this type of application has not seen even approximate empirical justification. Nevertheless, the principles upon which the methods are based are so obvious, and the potential applications so useful, that their presentation as preliminary engineering techniques seems worthwhile.

Furthermore, a glance at Chapters 3 to 6 reveals a host of additional mechanisms that were not approached in this chapter. The use of these other mechanisms in engineering applications awaits future development.

12

Critical research directions for the future

The foregoing chapters have summarized some of the research directions currently underway in intercropping research and have indicated some possible simple extensions of relatively routine ecological theory into the realm of intercropping research. The ideas expressed have ranged from the mainly intuitive to the highly quantitative engineering designs of intercropping systems, and have utilized some simple applications of set theory and some not so simple applications of partial differential equations. By and large to this point all material has been well-known on two levels, the level of the ecological theory to be applied and the level of the need for it in intercropping.

In this chapter we briefly consider several topics that are either not so well known as ecological formulations and/or not usually part of the typical research agenda of intercropping researchers. These topics are included here because of what appears to be a great need for their consideration among researchers. Because they are topics more for the future than the present, their presentation will be far more tentative and speculative than the rest of the material in the book has been.

Dynamic plant growth and interaction models

A situation in which the need for a dynamic approach is essential was presented in Chapter 10. The use of a static yield–density approach when applied to a system of tomatoes and beans resulted in excellent predictions some of the time. But it is of interest to examine the cases in which the predictions were far off the mark. Indeed in those experiments, of a possible 16 predictions, only 9 cases actually produced a prediction. That is in 7 of the 16 cases the model exploded (Vandermeer, 1986). A close examination of those situations revealed that if the competitive environment was sufficiently strong, the model predicted that some individuals in the population would die, a phenomenon usually referred to as self-thinning. Because of the linear nature of the model, it was even possible to predict individuals with negative yields, clearly a mathematical artifact generated by the linearity of the model.

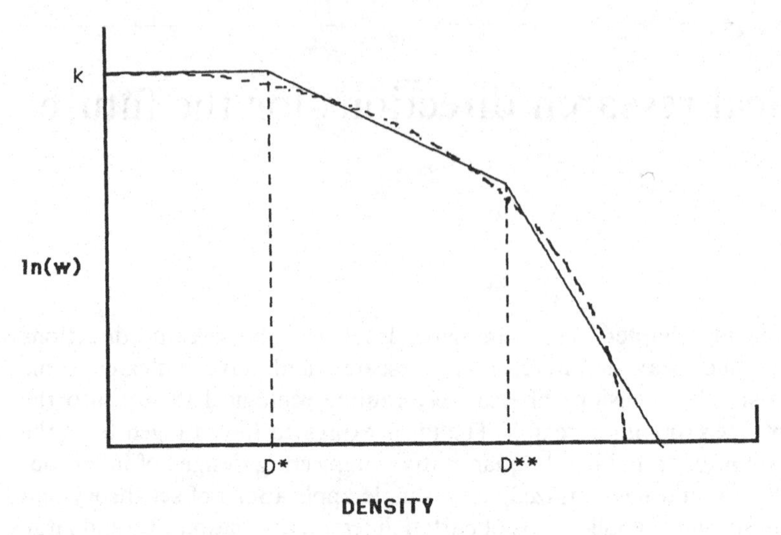

Fig. 12.1. The relationship between the biomass of an individual plant (w) and the density of the population in an even-aged stand.

Furthermore, the competitive effect that would have been exerted by individuals that later died made it impossible to simply recompute the yields after eliminating those individuals that experience mortality.

The generalities of yield and mortality relationships with regard to density, already discussed in Chapter 10, are pictured in Figure 12.1. The process can be thought of as three linear segments. First, densities are so low that each plant in the population grows unencumbered (between 0 and D^* in Figure 12.1). Second, yield is reduced by competition but there is no mortality (between D^* and D^{**}). Finally, competition is so intense that mortality (thinning) occurs (above D^{**}). While the actual relationship is likely to be a smooth monotonic function (e.g. Slatkin & Anderson, 1984), it is convenient to draw it in three linear segments so as to emphasize the three separate processes involved. In Figure 12.1 a dotted curve is included, indicating the possible true functional relationship, of which the three linear segments are only approximations. Since interplant distance, Δ, is simply the inverse square root of D, Figure 12.1 could easily be redrawn with $\ln\Delta$ as the abscissa, or for that matter simply Δ, as has been done in Figure 12.2. Taking this abstract line one step further, we note that the relative depression from k to w at a particular Δ is a reflection of competition at that Δ. Thus we can represent the same phenomena as a graph of the relative depression in biomass (the competition coefficient) as a function of Δ. Such a representation is presented in Figure 12.3. Conceptually, Figures 12.1, 12.2, and 12.3 are representations of the same phenomena.

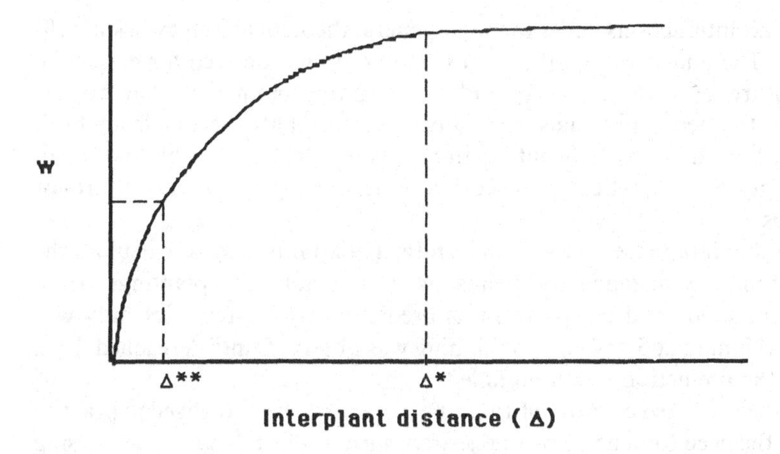

Fig. 12.2. The relationship between the biomass of an individual plant (w) and the distance to its competitors, in an even-aged stand.

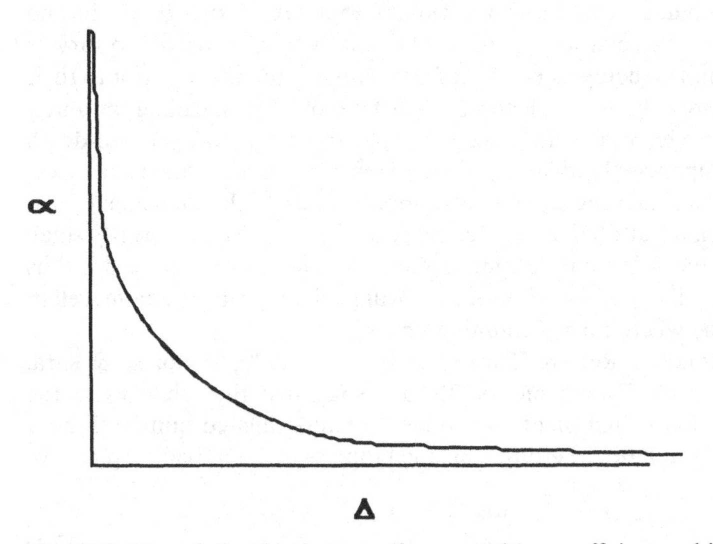

Fig. 12.3. The relationship between the competition coefficient and interplant distance.

Thinning occurs when the interplant distance becomes sufficiently small, that is, when Δ is equal to or less than Δ** in Figure 12.2. When using simple linear equations (as the example in Chapter 10), the thinning process appears as a negative biomass in the solution to the similtaneous equations. The quantity Δ** can thus be thought of as the 'critical thinning distance', and is easily computed in any given case (Vandermeer, 1986; and Chapter 10). A conceptually important issue here is the inclusion of thinning as a result of interspecific interactions under the same general theoretical framework as self-thinning. The analytical approach of Chapter 10 was derived for either the monoculture or diculture case, and the negative biomasses that appear represent thinned individuals generally. The mortality occurs from both intraspecific and interspecific interactions, casting the problem of thinning in the same general theoretical framework whether dealing with monocultures or dicultures.

From the data presented in Vandermeer (1986) it is easy to calculate the critical thinning distance for beans as 4.8 cm, for row plantings. In a subsequent series of density trials this prediction was tested. Not only was there no thinning at 5 or 4 cm, no thinning was observed until Δ reached 2 cm. Clearly, the prediction of 4.8 cm failed.

That the predicted critical self-thinning distance was not realized in practise suggests the need for a new line of research, specifically a dynamic one. Using simple yield equations restricts the possible biomass calculations to either the direct solution of the equations (which includes negative biomasses if thinning occurs) or elimination of those individuals that are predicted to die and recalculation of the equations using only the individuals expected to survive (which presumes no competitive effect from thinned individuals, prior to their death). Both procedures are clearly unrealistic biologically, calling for a new approach, one which takes into account both the death and prior-to-death effects of the suppressed individuals. What is required is a dynamic approach.

I hasten to add that the simple static approach, using the elementary yield equations, remains attractive in those cases where density is sufficiently small to allow yield reduction but not thinning (e.g. Vandermeer, 1986), and is thus not to be generally abandoned. Indeed, it will probably perform quite well in most situations where no self-thinning occurs.

Many authors (e.g. Reed & Holland, 1919; Reed, 1920; Shinozaki & Kira, 1956; Lotka, 1926; Robertson, 1923) have suggested that changes in the biomass of an individual plant over time are approximated quite well by a logistic-type formulation. Letting w be the biomass of an individual plant we may write

$$\frac{dw}{wdt} = r\frac{k-w}{k}, \tag{12.1}$$

where k is the 'unencumbered maximum biomass', which is to say the largest biomass that an individual plant will eventually attain when it is unencumbered by competition from other individuals, and r is the rate of growth before the biomass-yield related feedback takes effect (Shinozaki & Kira, 1956). Other functions have also been suggested including the monomolecular (Gregory, 1928); Gompertz (Gompertz, 1825; Amer & Williams, 1957), and Von Bertalanfy (Von Bertalanfy, 1957), usually in consideration of the growth of a plant part, rather than the whole plant.

More recently, plant growth analysts have been using the Richards function (Richards, 1959; Caustin & Venus, 1981), a generalized function which includes the monomolecular, logistic, Gompertz, and Von Bertalanfy function as special limiting cases (Caustin & Venus, 1981). While much criticism has been leveled at the analysis of whole plant growth using these equations generally (e.g. Caustin & Venus, 1981), such criticism has applied to the inability of such an approach to represent the details of an individual plant's growth trajectory, and the nonmechanistic philosophy involved with agglomerating all interesting physiological properties. The use to which I propose to put these equations is to build up from them, from the individual to the population. I thus maintain that much of the criticism that is relevant for the plant physiologist is probably irrelevant here.

Let the growth of a single individual be

$$\frac{dw}{w dt} = r\left[1 - \left(\frac{w}{k}\right)^b\right], \tag{12.2}$$

where r is the unencumbered growth rate, k is the unencumbered final biomass, b is the 'rate of feedback' (here we refer to w/k as the feedback), and w is the biomass of the plant at some point in time.

Adding a second individual, we can expect the rate of growth of the first to be reduced, roughly proportional to the biomass of the second, at any particular point of time. In particular, r is not expected to change since it is the limiting growth rate. Nor is k expected to change since it is the theoretically largest biomass that an individual could attain. Nor is b expected to change, since it is the rate at which the feedback effect changes. But we do expect the feedback function itself to change. That is, if the original growth rate responded to a function of the relative attainment of maximum biomass, w/k, we should expect the same thing to happen when the second individual is in the vicinity. In essence, the tissues of the first species respond to nutrient depletion, or shading, or some other form of competition, regardless of its source. Thus the w in the w/k function could more generally be thought of as 'competing tissue'. If that is the case, the biomass of the second individual represents part of that competing tissue, although not in a one-to-one correspondence to the biomass of the first. Thus we think of the second individual as effectively augmenting the extant biomass to which the growth rate of the first must respond. The feedback function then becomes, $(x + \alpha y)/k$, where y is the biomass of the second individual and α is a scaling factor that converts biomass of the second into equivalents of the first.

A variety of factors affect α to the extent that the plant in question is determinate and resources are abundant, α will tend to zero. This fact stems from the simple idea that two determinate plants growing in a non-depletable environment will simply not notice one another. The opposite situation is two indeterminate plants growing in a resource-poor environment, in which case α will be large. Similarly, the physical proximity of the two plants will affect the value of α. If the plants are far enough removed that they have little effect on one another, α obviously will be near zero.

With the above assertions we can reformulate equation (11.2) to describe the simultaneous growth of the two individuals as

$$\frac{dw}{wdt}=r\left[1-\left(\frac{w-\alpha y^b}{k}\right)\right], \tag{12.3a}$$

$$\frac{dy}{ydt}=r\left[1-\left(\frac{y-\alpha w}{k}\right)^b\right]. \tag{12.3b}$$

The equilibrium form of these equations is

$$\left.\begin{aligned} k&=w+\alpha y,\\ k&=\alpha w+y,\end{aligned}\right\} \tag{12.4}$$

and the equilibrium value for the state variables is

$$w=y=\frac{1-\alpha}{1-\alpha^2}k. \tag{12.5}$$

Note that since the exponent b effectively cancels out of the equilibrium form, that form will be identical for the monomolecular, logistic, Gompertz, Von Bertalanfy, or Richards form of the equations. The approach to that equilibrium position will, of course, not be the same for all of these functions, nor will the potentialities for more complex dynamic behavior.

Dynamically, the concern is with either a stable equilibrium in which both individuals approach the value stipulated by equation (12.5), or an unstable equilibrium in which one individual approaches k and the other approaches zero, which is to say one dominates and the other is suppressed. Dominance and suppression will occur whenever $\alpha > 1$ (since the α and k are constant in both equations, the usual criteria on the eigenvalues reduce to this simple relationship).

The use of equations (12.3) to describe the dynamics of competition has not yet been presented in the literature. A first attempt, using potted common beans, is presented in Figure 12.4. The results are encouraging. It is worth noting that since the other growth equations (monomolecular, logistic, Gompertz, Von Bertalanfy) are all special cases of the Richards function, any analytical results for these equations will apply for the other growth functions as well.

An initial glimpse of some analytical details can be gained by a simple two dimensional qualitative analysis. As readily observable in Figure 12.5, two alternative configurations are possible, a stable equilibrium (Figure 12.5(*a*)) in which both individual plants converge on the value stipulated in equation

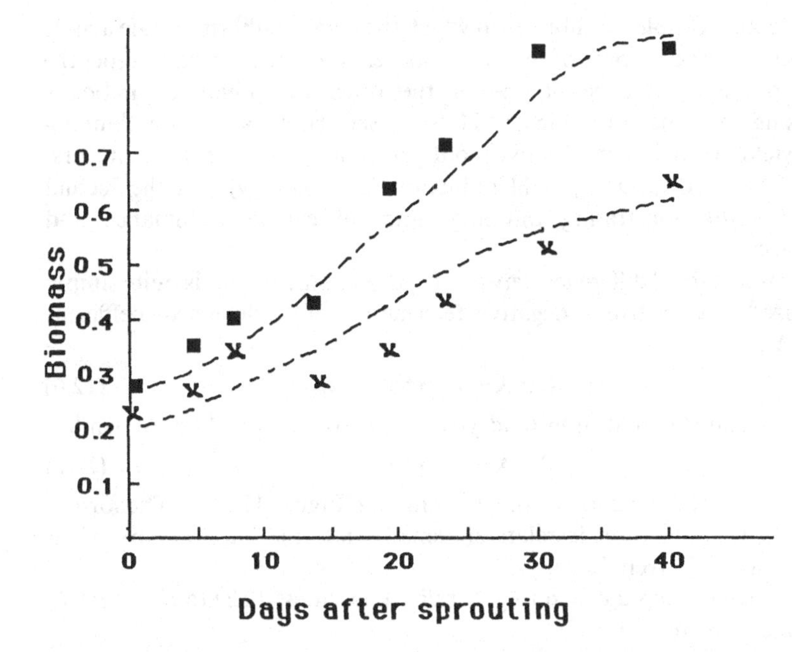

Fig. 12.4. Growth equations (dashed lines) and the growth of potted *Phaseolus vulgaris*. Squares represent growth of an individual plant, while ×s represent the growth of an individual plant in competition with another.

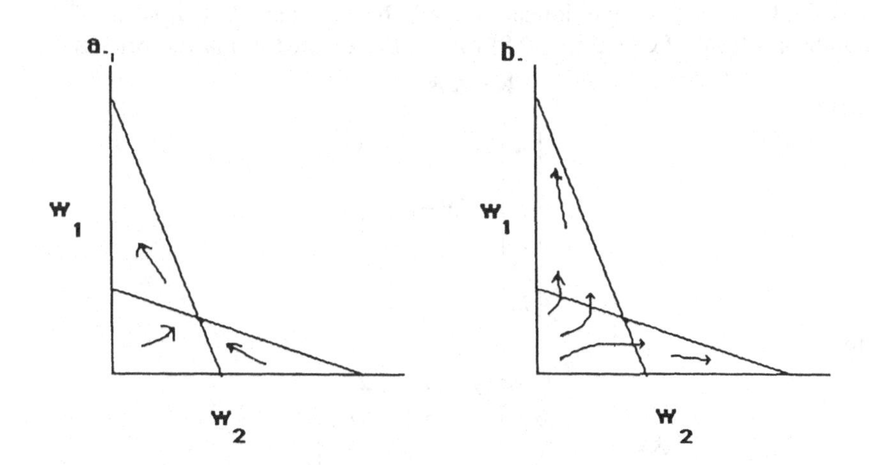

Fig. 12.5. The two qualitatively different outcomes of two model plants in competition: (*a*) stable equilibrium in which both plants will coexist; (*b*) unstable equilibrium in which one individual will dominate and the other be suppressed.

(12.5), and an unstable equilibrium in which the ideal equilibrium state acts as an attractor over only part of the phase-space, eventually acting to repel the system to the dominance of one or the other individual, theoretically suppressing the other to its death. Thus we see the metaphor for thinning and/or yield reduction in the two configurations of Figure 12.5, the first (Figure 12.5(*a*)) illustrating yield reduction (or plasticity) and the second (Figure 12.5(*b*)) illustrating thinning (and, of course, dominance and suppression).

The mathematical difference between Figure 12.5(*a*) and (*b*) is quite simply summarized as a positive or negative determinant of the detached coefficient matrix (**A**), that is,

$$\text{Det}\,\mathbf{A} = 1 - \alpha^2 > 0 \tag{12.6}$$

stipulates a stable equilibrium (and yield reduction, Figure 12.5(*a*)), while

$$\text{Det}\,\mathbf{A} = 1 - \alpha^2 < 0 \tag{12.7}$$

stipulates an unstable equilibrium (the thinning, Figure 12.5(*b*)). The former occurs when competition is relatively small while the latter occurs when competition is relatively large.

It is a small conceptual step to generalize equations (12.3) to represent N individuals, namely,

$$\frac{\mathrm{d}w_i}{w_i \mathrm{d}t} = r\left\{1 - \left[\frac{w_i - \sum_{j=i} \alpha_{ij} w_j}{k}\right]^b\right\}, \tag{12.8}$$

where w_i is the biomass of the ith individual, and α_{ij} is the effect of the jth individual on the ith. No longer are all the αs equal, but $\alpha_{ij} = \alpha_{ji}$. The equilibrium form of equation (12.8) can be represented in matrix form as

$$\mathbf{K} = \mathbf{AW},$$

where

$$\mathbf{K} = \begin{vmatrix} k \\ k \\ \cdot \\ \cdot \\ \cdot \\ k \end{vmatrix}, \quad \mathbf{W} = \begin{vmatrix} w_1 \\ w_2 \\ \cdot \\ \cdot \\ \cdot \\ w_n \end{vmatrix},$$

and

$$\mathbf{A} = \begin{vmatrix} 1 & \alpha_{12} & \alpha_{13} & \cdots & \alpha_{1n} \\ \alpha_{21} & 1 & \alpha_{23} & & \alpha_{2n} \\ \cdot & & & & \\ \cdot & & & & \\ \cdot & & & & \\ \alpha_{n1} & & & & 1 \end{vmatrix}$$

Thus we have a multidimensional analogue to the simple two-individual analysis. It is also true (recall that the matrix **A** is symmetrical) that the determinant of the detached coefficient matrix will stipulate if the entire system has a stable equilibrium point or not, and thus whether or not thinning is expected. Thus, during the process of growth, if an even-aged stand is sufficiently dense to require thinning, $\det \mathbf{A} < 0$. The individuals in the population will grow until the most suppressed individual dies. At that point Det **A** changes (since one of the individuals is no longer there), presumably, though not necessarily, becoming larger. If Det **A** is still less than zero, another individual will die. This process continues until Det $\mathbf{A} > 0$, at which point all of the individuals survive, their biomasses given as

$$\mathbf{W} = \mathbf{A}^{-1}\mathbf{K}. \tag{12.9}$$

Equation (12.9) should be usable for computation of the individual biomasses of the thinned population.

But while the population is undergoing self-thinning, use of equation (12.9) to compute biomasses would not be appropriate since one would actually be computing at least one equilibrium biomass that is wrong – the biomass for the individual that is on its way to suppression (i.e. if we were to compute the biomasses at the intersection of the isoclines in Figure 12.5(*b*), we would not be computing the true values, which would be one biomass near zero and the other near k).

During the process of self-thinning, any two individuals can be viewed as interacting on a graph of w_1 vs w_2. If only the two of them were in the population they would behave according to their isocline structure (i.e. as in Figure 12.5). But because other individuals are in the population, those two will not behave as a simple competing pair. This process is illustrated in Figure 12.6. The hypothetical population contains five individuals. Individual 3 receives strong competitive effects from individuals 1 and 4, and is thus the first individual to be suppressed. If we plot the biomass of individual 2 against individual 1, we can qualitatively picture dynamic changes in these two individuals during the process of thinning. Overall, there will be a tendency to approach the 45-degree line, even while the population is in the process of self-thinning. Since that 45-degree line generally represents the locus of all the equilibrium points of all the pairs of individuals in the population, many of which are unstable, we can view any two nonsuppressed individuals as tending towards this unstable locus.

Extending this approach to two species is a simple matter, using the biological reasoning developed above. The feedback component in the monoculture is

$$\frac{w_i - \sum_{j=i} \alpha_{ij} w_j}{k}. \tag{12.10}$$

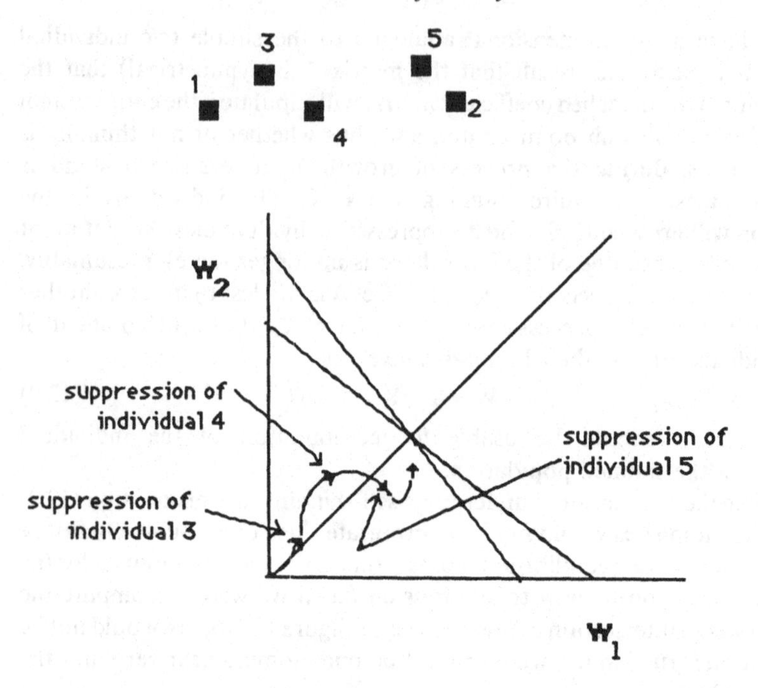

Fig. 12.6. Likely trajectories for individuals 1 and 2 when embedded in a population planted in a pattern corresponding to the diagram above the graph.

Following the reasoning developed earlier, that the relative growth rate is diminished by the effective competitive tissue, individuals of a second species should have the same quantitative effect as those of the first species. Thus we only need to expend the feedback component (equation (12.10)) to include the individuals of the second species, scaling them with a different coefficient. Thus the diculture equivalent of the system represented by equation (12.8) is

$$\frac{dw_i}{w_i dt} = r\left\{1 - \left[\frac{w_i - \sum_{j=i} \alpha_{ij} w_j - \sum_{j=i} \beta_{ij} y_j}{k}\right]^b\right\}. \tag{12.11}$$

Equation (12.11) will have an equilibrium form as presented earlier. As dynamic equations, they have not yet been studied either theoretically nor empirically.

In development of the present dynamic approach, at least four tasks remain. First, the form of the equations must be studied and verified in the intraspecific case. Initial data (Figure 12.4(*a*) and (*b*)) are encouraging, but it represents only a small data base. Growth of individual, pairs, and triplets needs to be investigated also. Second, the form of the equations must be studied and verified in the interspecific case. Third, the form of the equations must be verified under field conditions. Finally, field plots need to be employed to verify that dynamic predictions from the model are correct.

This dynamic approach, after considerable development and verification, would be useful in the same sense as the approach of Chapter 10, with the additional utility of applying to a larger variety of circumstances, particularly those involving thinning or those involving different dates of planting and harvesting of the different crop species in an intercrop. Rotation intercrops and combinations of perennials and annuals are two obvious situations that come to mind. Furthermore, the incorporation of other topics discussed earlier might be simple extensions. The dynamics of weed growth in intercrops will be the simple addition of a third species to the system with little in the way of conceptual addition. Various weed control strategies might then be investigated in the model system.

The question of mechanization

The practise of intercropping in temperate climates became obsolete when machines such as McCormick's reaper came into extensive use. Such machines were very primitive by modern standards, and were really nothing more than a mechanical scythe. The later combination of threshing and harvesting into one operation, the modern 'combine' was really quite a sophisticated jump in technology, but its focal point was at the interface of harvesting and processing – seed heads were harvested and then the seeds automatically removed from the head. This sort of machine has become something of a paradigm. A large combine working day and night through fields of uninterrupted grain, each plant exactly the same height as its neighbor, the operator comfortable in the air-conditioned cab, enjoying the latest hit songs in the stereo tape deck. The ultramodern combine, combined with the hyperimproved grain, forms the paradigm of technified agriculture as much in the Soviet Union as in the United States. Such a vision extends also to the Third World, where it remains more-or-less yet a vision.

It is clear that with such a vision, intercropping remains a technique for the premodern farm, perhaps of use for peasant agriculturalists on the hillsides of Guatemala, but hardly what anyone would imagine as part of the ultramodern farm. Indeed, the requirements of mechanical harvesting and seeding are popularly thought to be overwhelming impediments to the use of intercropping. To paraphrase a reviewer of one of my grant proposals, '. . . intercropping has no place in modern agriculture since you cannot mechanically harvest an intercrop'. Remarkably, I have heard the same objection to intercropping in the underdeveloped world in systems for which no mechanical harvester exists in the entire country – 'Intercrop tomatoes with beans? How will you harvest (read mechanically) the tomatoes?' So intense is this vision of modern agriculture that merely the anticipation of mechanical harvesters in the far distant future is apparently enough to promote monocultures.

A possible alternative to this vision is provided by some proponents of alternative or appropriate technology. For a variety of reasons they argue that mechanical harvesting itself is not a desirable goal, and that techniques that do not require it are good techniques for that reason alone. I do not hold this view. I accept the vision of mechanical harvesting (and seeding) as a vision appropriate for the future in the Third World and as a reality of agriculture in the developed world. But I would argue that this vision is not what it appears to be with regard to intercrops, that mechanical harvesting is quite compatible with production in intercrops.

Consider the development of the mechanical tomato harvester (Friedland & Barton, 1975; Vandermeer, 1986*b*; Rosset & Vandermeer, 1986; Dickman, 1978). When the notion of mechanically harvesting tomatoes occurred to a visionary technician in California in the early 1940s, he became something of the brunt of jokes. Since tomatoes were soft fruits that ripened continually, and since workers harvested a single field repeatedly, the idea that one could mechanize this operation was regarded as close to lunacy. For a variety of social and economic reasons the technology was nevertheless pursued for over 15 years, and finally in 1958 the first mechanical tomato harvester appeared in the tomato fields of California. A short six years later, over 90% of California's processing tomatoes were being mechanically harvested – with a machine that 20 years earlier was regarded as the unfeasible dream of an unrealistic visionary.

The point to be emphasized here is that a machine that can mechanically harvest and thresh a field of corn and beans simultaneously is probably quite a bit easier to imagine now than was a machine that could mechanically harvest tomatoes in 1945. It is simply not true that intercrops cannot be mechanically harvested. Now that we have what are virtually robots roaming the tomato fields of California, not only mechanically harvesting the fruits but automatically sorting them with electric eyes, I suggest that the lunatic view is that we cannot do the same thing for intercrops, where the technical problems are frequently far simpler.

Indeed the technical limitations are arguably quite trivial. Probably the major impediment to mechanically harvesting intercrops is an economic one (Erbach & Lovely, 1976). If the general political and economic climate of a country dictates a cheap labor force, it probably won't make economic sense to develop machines to replace labor. Since mechanization is in principle designed to maximize production per farmer or per worker rather than per hectare, a poorly paid labor force is a definite stumbling block to the development of mechanical harvesters. However, as labor costs rise, or as they are anticipated to rise (as they were in the 1940s in California, Dickman, 1978), the economic feasibility of mechanical harvesting rises. The regions or countries that anticipate these changes first are likely to gain considerable advantage in world markets.

A variety of special circumstances are presented with intercropping. In some situations it will be the case that a later ripening crop can be harvested mechanically after the earlier ripening crop has been harvested by hand or, if simultaneously ripening, one crop is hand harvested first and the other is machine harvested. The only problem associated with this technique is the same problem faced by producers when mechanically harvesting a weedy field, clearly a problem, but not insurmountable. A similar situation occurs in the case when a second crop is intercropped because it offers an advantage to the first crop only, and is of no harvestable value itself, as in the case of simple cover crops such as clover. Here again, it is the problem of harvesting a weedy field.

When the intercrop is a strip intercrop, machine adaptations are simply questions of size. The same harvesting principles are involved as with monocultures, but the machine must be small enough to accommodate narrow strips. As the width of the strip becomes smaller, we naturally grade into the alternate row intercrop, in which case it will rarely be possible to have a machine that harvests a single row. Here the problem becomes one of changing harvesting technology from row to row on the same machine. We can imagine cases where this might be simply a matter of adjusting the size of a cutter depending on which row it is cutting, while in other cases the entire technology of cutting might be different for the two crops. In either case the conceptual problems are trivial, despite the fact that the engineering problems may be great.

Finally, there are cases where a great deal of imagination will have to go into mechanical harvesting. Here we can only point to the successes already realized by first world engineers in the development of what are arguably robots that harvest. From a technical point of view it is only a lack of imagination that stands in the way of harvesting whatever intercrop mechanically. As Sanchez (1976, p. 511) has noted: 'In view of the development of machines for harvesting tobacco, tomatoes, and grapes, it appears unduly pessimistic to assume that intercropped systems too cannot be fully mechanized.' The social and economic constraints are, of course, somewhat more difficult to overcome.

The question of genetic improvement

With a few exceptions, the vast majority of intercropping research is carried out with modern improved varieties. These varieties are invariably selected for performance under monocultural conditions and the wonder is not that intercrops do well, but rather that they do so well with crop varieties that have been selected to perform under a totally different set of conditions. For example, in Nigeria a local tomato cultivar performed much better than an improved variety when intercropped with okra (Olosantan, 1985). Even with modern varieties there is enough accumulated evidence on variable performance in intercrops of different genotypes to assure that significant genetic material is available for a breeding program (e.g. Rao & Willey, 1983; Francis, 1986; Francis *et al.*, 1978; Maken & Doto, 1982; Brakke *et al.*, 1983; Kwano & Thung, 1982). In general, what might intercrop advantage look like if we had varieties that had been improved for the last 30 years under intercropping situations?

There seems to be some fundamental misunderstanding of what genetic improvement in intercrops should be about. For example, 'An agriculturalist would view complementation (i.e. intercrop overyielding) based on differing root behavior as evidence of defective phenotypes and would look for a cultivar or species capable of exploiting the full profile.' (Loomis, 1984.) The absurdity of such a position is most easily seen with reference to perennial/annual combinations. Since young oil palms are not capable of mining the full soil profile, Loomis's agriculturist would presumably look for oil palm cultivars that could mine nitrogen as efficiently as the oil palm–legume combination normally used!

Improvement programs might be envisioned in two ways – as exploiting reduced competition mechanism more efficiently or as improving a facilitative mechanism.

Some tentative guides to intercrop improvement through reducing interspecific competition can be gleaned from the rich literature on theoretical evolutionary ecology (for excellent primers see either Pianka, 1976; or Roughgarden, 1972). Of special importance is the classic question posed by MacArthur & Levins (1967), which asked how similar can two species be in their niche requirements before they compete strongly enough so as not to be able to coexist. The idea is really intuitive. Consider, for example, two bird species, one of which forages towards the tops of trees and the other of which forages near the bottoms of trees. Assuming that the two are utilizing the same resource, we can easily imagine that there would be coevolutionary pressure against those who foraged at intermediate levels, for they would experience the most intense interspecific competition. We thus expect coevolutionary pressure to separate the feeding niches, resulting in one species that specializes on feeding on the tops of trees and the other species that specializes on feeding near the bottoms of the crowns. But if the separation becomes large enough, it could result in no foraging behavior near the center of the crown, which might enable another species to enter the community, thus forcing the first two into a new round of competitive interaction, this time with a third species. We can thus conceive of a sort of balance of coevolutionary pressure, on the one hand separating the niches, on the other hand maintaining a certain degree of overlap.

In those cases where the competitive production principle operates (or will operate in the improved varieties) this idea and its varied derivatives (e.g. Roughgarden, 1972; Levins, 1968) might be put to good use in planning a breeding program. What natural selection has been assumed to do may provide the breeder with a rough guide as to what artificial selection should do. The underlying framework already set up in Chapter 3 forms the static base, and the selective regime forms the dynamic superstructure. In short, traditional improvement programs concentrate on the improvement of k in all microhabitats. The new programs must concentrate on the encumbered rather than the unencumbered yields in all microhabitats. This means that not only should we have the goal of increasing the k, but also decreasing the interspecific competitive effect.

As in the case of traditional improvement programs, simply mimicking the action of natural selection, choosing new material based on the performance of individuals, will likely not produce rapid results. Rather, understanding the mechanism of yield has in the past been a key to the development of breeding programs, such that particular structures or processes could be attacked by the artificial selectionist to increase yields indirectly. Similarly, in developing improvement programs for intercropping systems we will need to know the mechanisms of competition so as to select against it efficiently. In many cases the genetic material is readily available. For example, Ayuk-Takem & Chedda (1985) found that early maturing varieties of maize with small erect leaves were best suited for intercropping with cocoyams, presumably because of a partitioning of the niche along both the time and light interception axes. The first steps towards development of an improvement program may thus be the elaboration of the detailed mechanisms of competition, where the theoretical outlines of Chapter 5 may become important.

The picture is dramatically different if the operating principle is facilitation. Again the development of an efficient breeding program depends on understanding mechanism, but here the goal is not necessarily to decrease the competitive effect but, rather, to increase the environmental modification. For example, if one of the crops is effectively acting as a trap-crop for a key pest, a program of increasing its susceptibility to oviposition of that pest may be the best plan.

References

Abalu, G.O.I. 1977. A note on crop mixtures in Northern Nigeria. *Samaru Res. Bull.* **276**:212–20.

Agamuthu, P. & W.J. Broughton. 1985. Nutrient cycling within the developing oilpalm-legume ecosystem. *Agriculture, Ecosystems and Environment* **13**:111–23.

Agboola, A.A. & A.A. Fayemi. 1971. Preliminary trials on the intercropping of maize with different tropical legumes in western Nigeria. *J. Agric. Sci., Camb.* **77**:219–25.

Aggaoili, L.B. 1961. Cacao in coconut plantations. *Coffee and Caco J.* **4**:223–9.

Aiyer, A.K.Y.N. 1949. Mixed cropping in India. *Indian J. Agr. Sci.* **19**:439–543.

Alas, L.N. 1974. Breve descripción del sistema de producción del pequeno productor en El Salvador. Apendice F. In *Conferencia sobre sistemas de producción agrícola para el trópico. Cent. Agron. Trop. de Invest. y Ensenanza.* Informe final. CATIE. Turrialba. Costa Rica.

Alcorn, J.B. 1984. *Huastec Mayan Ethnobotany*. University of Texas Press, Austin, Texas.

Allen, J.R. & R.K. Obura. 1983. Yield of corn, cowpea and soybean under different intercropping systems. *Agron. J.* **75**:1005–9.

Altieri, M.A., A. van Schoonhoven & J.D. Doll. 1977. The ecological role of weeds in insect pest manatement systems: a review illustrated with bean (Phaseolus vulgaris L.) cropping systems. *PANS* **23**:195–205.

Amer, F.A. & W.T. Williams. 1957. Leaf growth in *Pelargonium zonale. Annals of Botany.* **21**:339–42.

Anderson, M.C. 1964. Light relations of terrestrial plant communities and their measurement. *Biol. Rev.* **39**:425–86.

Anderson, M.C. 1966. Some problems of the simple characterization of the light climate in plant communities. In R. Bainbridge, G.C. Evans and O. Rackham (eds.) *Light as an Ecological Factor. Symp. Br. Ecol. Soc.* 77–90.

Andrews, D.J. & A.H. Kassam. 1976. Importance of multiple cropping in increasing world food supplies. pp. 1–10, In R.I. Papendick, A. Sanchez, and G.B. Triplett (eds.) *Multiple Cropping.* ASA (American Society of Agronomy) Spec. Pub. 27.

Andrews, R.E. & E.I. Newman. 1970. Root density and competition for nutrients. *Oecol. Plant. Gauthier-Villars* **5**:319–34.

Aranguren, J., G. Escalante, & R. Herrera. 1982*a*. Nitrogen cycle of tropical perennial crops under shade trees. I. Coffee. *Plant and Soil* **67**:259–69.

Aranguren, J., G. Escalante, & R. Herrera. 1982*b*. Nitrogen cycle of tropical perennial crops under shade trees. II. Cacao. *Plant and Soil* **67**:259–69.

Arnon, I. 1972. *Crop production in Dry Regions.* Leonard Hill, London.

Arny, A.C., T.E. Stoa, C. McKee & A.C. Dillman,. 1929. *Flax Cropping in Mixture with Wheat, Oats, and Barley*. Tec. Bull. 133. US Dept. of Agric., Washington DC, USA.

Ashokan, P.K., V. Nair, & K. Sudhakara. 1985. Studies on cassava-legume intercropping systems for the Oxisols of Kerala State, India. *Trop. Agric.* (*Trinidad*) **62**:313–18.

Aspinall, D. 1960. An analysis of competition between barley and white presicaria. II. Factors determining the course of competition. *Ann. Appl. Biol.* **48**:637–54.

Atwal, A.S. & A. Mangar. 1969. Repellent action of root exudates of *Sesamum orientale* against the root-knot nematode, *Meloidogyne icognita* (Heteroderidae: Nematoda). *Indian J. Entomol.* **31**:286.

Ayeni, A.O., W.B. Duke, & I.O. Akobundu. 1984. Weed interference in maize, cowpea and maize/cowpea intercrop in a subhumid tropical environment. I. Influence of cropping season. *Weed Res.* **24**:269–79.

Ayuk-Takem, J.A. & H.R. Chedda. 1985. Grain yield potential of some diverse maize (*Zea mays* L.) morphotypes intercropped with cocoyam (*Xanthosoma sagittifolium*) *Exp. Agric.* **21**:145–52.

Bach, C.E. 1980*a*. Effects of plant density and diversity on the population dynamics of a specialist herbivore, the striped cucumber beetle, *Acalymma vittata* (Fab.). *Ecology* **61**:1515–30.

Bach, C.E. 1980*b*. Effects of plant diversity and time of colonization on an herbivore-plant interaction. *Oecologia* **44**:319–26.

Bach, C.E. 1981. Host plant growth form and diversity: Effects on abundance and feeding preference of a specialist herbivore, *Acalymma vittata* (Coleoptera: Chrysomelidae). *Oecologia* **50**:370–5.

Bailey, C.H. 1914. The composition and quality of wheat grown in mixtures with oats. *Agron. J.* **6**:204–10.

Baker, E.F.I. 1978. Mixed cropping in Northern Nigeria. I. Cereals and groundnuts. *Expl. Agric.* **14**:293–8.

Banks, H.T., P.M. Karieva & K.A. Murphy. 1987. The ladybird/aphid story, or how all the 'lady's' gather at the site of the 'goodies'. MS.

Bantilan, R.T., M. Palada & R.R. Harwood. 1974. Integrated weed management. 1: Key factors affecting weed/crop balance. *Philippine Weed Sci. Bull.* **1**:14–36.

Barber, S.A. 1962. A diffusion and mass-flow concept of soil nutrient availability. *Soil Sci.* **93**:39–49.

Barley, K.P. 1970. The configuration of the root system in relation to nutrient uptake. *Adv. Agron.* **22**:159–201.

Bebawi, E.F. & A.H. Abdelaziz. 1983. Grain sorghum responses to pure stands and mixtures under irrigation. *Trop. Agric.* (*Trinidad*) **60**:262–4.

Bebbington, A.G. & W. Allan. 1933. Progress report, Empire Cotton Growing Corporation Experiment Stations Northern Rhodesia, season 1931–32.

Beets, W.C. 1982. *Multiple Cropping and Tropical Farming Systems.* Westview Press, Boulder, 156 pp.

Begonia, G.B. & B.L. Mercado. 1974 Evaluation of herbicides for weed control in cabbage-tomato intercropping system. In *Weed Science Report* 1973–74. Laguna, Philippines: Dept. of Agronomy, University of Philippines. pp. 38–40.

Bergelson, J. & P. Kareiva. 1987. Barriers to movement and the response of herbivores to alternative cropping patterns. *Oecologia* **71**:457–60.

Best, 1962. The evaluation of Panman's natural evaporation formula by electronic computer. *Aust. J. Ppl. Sci.* **15**:61–4.

Bhatnagar, V.S. & J.C. Davies. 1969. Pest management in intercrop subsistence farming. In Willey, R.W. (ed.) *Proceedings of the International Workshop on Intercropping,* ICRISAT, Hyderabad, India. pp. 249–57.

Bhoj, R.L. & P.C. Kapoor. 1970. Intercropping of maize in spring planted sugarcane gives high profits with adequate nitrogen use. *Indian J. Agron.* **15**: 242–6.

Black, C.C., T.M. Chen & R.H. Brown. 1969. Biochemical basis for plant competition. *Weed Sci.* **17**:338–44.

Bleasdale, J.K.A. & J.A. Nelder. 1960. Plant population and crop yield. *Nature* **188**:342.

Blencowe, J.W. 1969. Castor: a prospective intercrop in Malayan plantations. In *Malayasian Oil Palm Conference,* 2nd, Kuala Lumpur. *Progress in Oil Palm.* pp. 238–51.

Bodade, V.N. 1964. Mixed cropping of groundnut and jowar. *Indian Oilseeds J.* **8**:297–301.

Böhning, R.H. & C.A. Burnside. 1956. The effect of light intensity on rate of apparent photosynthesis in leaves of sun and shade plants. *Am. J. Bot.* **43**:557–61.

Boucher, D.H. 1985. *The Biology of Mutualism: Ecology and Evolution.* Croom Helm, London, 388 pp.

Bourdeau, P. 1954. Oak seedling ecology determining segregation of species in piedmont oak-hickory forests. *Ecological Monographs* **24**:297–320.

Bray, R.H. 1954. A nutrient mobility concept of soil-plant relationships. *Soil Sci.* **78**:9–22.

Brakke, J.P., C.A. Francis, L.A. Nelson & C.O. Gardner. 1983. Genotype by cropping system interactions in maize grown in a short season environment. *Crop Science* **23**:868–70.

Brown, C.M. & D.W. Graffis. 1976. Intercropping soybeans and sorghum in oats. *Illinois Res.* **18**:3–4.

Budowski, G. 1981. The place of agroforestry in managing tropical forests. In Mergen, F. (ed.) *Proc. Intern. Symp. on Trop. Forests: Utilization and Conservation,* Yale University, New Haven, USA. pp. 181–94.

Bussell, F.P. 1937. Oats and barley on New York farms. *Cornell Ext. Bull.* 376.

Camlin, M.S., T.J. Gilliland & R.H. Stewart. 1983. Productivity of mixtures of Italian ryegrass (*Lolium multiflorum* cultivar RvP) and red clover (*Trifolium repens*). *Grass Forage Sci.* **38**:73–8.

Carvalho, M. de. 1969. O algodao como cultura intercalar do sisal. *Revista Agricola, Mozambique,* **11**:7–8.

Case, T.J. & M.E. Gillpin. 1974. Interference competition and niche theory. *Proc. Nat. Acad. Sci.* **71**:3073–7.

Castillo, M.B., M.S. Alejar & R.R. Harwood. 1976. Nematodes in cropping patterns. II. Control of *Meloidogyne incognita* through cropping patterns and cultural practices. *Philippine Agriculturist* **59**:295–312.

Castin, E.M., S. SanAnntonio & K. Moody. 1976. The effect of different weed control practices on crop yield and weed weight in sole cropped and intercropped corn and mung beans. 7th Ann. Conf. of Pest Control Council, Philippines. 5–7 Mau. Cagauan de Oro Citu. Philippines.

Caustin, D.R. & J.C. Venus. 1981. *The Biometry of Plant Growth.* Edward Arnold, 307 pp.

Cecilio, P. & D.P. Janos. 1987. Vesicular-Arbuscular mycorrhizas cause Maize-Solanum Polyculture overyielding.

Chang, H. & R.C. Lin. 1960. Studies on the interplanting of sugar cane with spring paddy rice. (in chinese.) *Report of the Taiwan Sugar Exp. St.* **22**:87–100.

Chang, J. 1968. *Climate and Agriculture: an Ecological Survey.* Aldin Pub. Co. Chicago.

Cheng, V.W. 1970. Improving the performance of catch crops in Malaysia. In Blencowe, J.W. and V.W. Blencowe (eds.) *Crop Diversification in Malaysia.* Kuala Lumpur. pp. 66–77.

Chiariello, N., J.C. Hickman, & H.A. Mooney. 1982. Endomycorrhizal role for interspecific transfer of phosphorous in a community of annual plants. *Science* **217**:941–3.

Clark, E.A. & C.A. Francis. 1985. Transgressive yielding in bean: maize intercrops; interference in time and space. *Field Crops Res.* **11**:37–53.

Clements, F.E. 1928. *Plant Succession and Indicators.* H.W. Wilson Co. New York.

Colwell, R.K. & E.R. Fuentes. 1975. Experimental studies of the niche. *Ann. Rev. Ecol. Syst.* **6**:281–310.

Colwell, R.K. & D.J. Futuyma. 1971. On the measurement of niche breadth and overlap. *Ecology* **52**:281–310.

Cordero, A. 1977. Principles of intercropping: Effects of nitrogen fertilization and row arrangement on growth, nitrogen accumulation, and yield of corn and interplanted understory annuals. Ph.D. Thesis NC State Raleigh.

Cowling, D.W. & D.R. Lockyer. 1967. A comparison of the reaction of different grass species of fertilizer nitrogen and to growth in association with white clover. I. Yield of nitrogen. *J. Br. Grassl. Soc.* **22**:53–61.

Cromartie, W.J. 1975. The effect of stand size and vegetational background on the colonization of cruciferous plants by herbivorous insects. *J. Appl. Ecol.* **12**:517–33.

Cromartie, W.J. 1980. The environmental control of insects using crop diversity. In Pimentel, D. (ed.) *Handbook of Pest Management.* Chemical Rubber Company Series in Agriculture, Boca Raton, Florida, USA.

Crookston, R.K. 1976. Intercropping: a new version of an old idea. *Crops and soils* **28**:7–9.

Culver, D.C. 1970. Analysis of simple cave communities: niche separation and species packing. *Ecology* **51**, 949–58.

Culver, D.C. 1985. *Cave Communities.* Harvard University Press.

Dalal, R.C. 1974. Effects of intercropping maize with pigeonpeas on grain and nutrient uptake. *Exp. Agr.* **10**:219–24.

Davis, J.H.C., M.C. Amézquita & J.E. Munoz. 1981. Border effects and optimum plot sizes for climbing beans. (*Phaseolus vulgaris*) and maize in association and monoculture. *Exp. Agr.* **17**:127–35.

Dempster, J.P. 1957. The population dynamics of the Moroccan locust (*Dociostauras maroccanus* Thunberg). *Anti-Locust Bull.* **27**:1–60.

Dempster, J.P. & T.H. Coaker, 1974. Diversification of crop ecosystems as a means of controlling pests. In Price Jones, D. and M.E. Solomon (eds.) *Biology in Pest and Disease Control.* Blackwell Scientific, Oxford. pp. 106–14.

Dickman, 1978. Interviews with persons involved in the development of the Mechanical Tomato harvester, the compatible processing tomato and the new agricultural systems that evolved. Davis: Shields Library, Oral History Center.

Donahue, R.L., R.W. Miller & J.C. Shickluna. 1977. *An Introduction to Soils and Plant Growth.* Prentice-Hall, New Jersey.

Donald, C.M. 1958. The interaction of competition for light and for nutrients. *Aust. J. Agric. Res.* **9**:421–35.

Donald, C.M. 1961. Competition for light in crops and pastures. in Milthorp, F.L. (ed.), *Mechanisms in Biological Competition. Symp. Soc. Exp. Biol.* **15**:283–313.

Edje, O.T. 1979. Cropping systems for small farmers. Bunda College for Ag., *Res. Bull.* **10**:10–33.

Egunjobi, O.A. 1984. Effects of intercropping maize with grain legumes and fertilizer treatment on populations of *Pratylenchus brachyurus* Todfrey (Nematoda) on the yield of maize (*Zea mays* L.) *Protection Ecol.* **6**:153–67.

Ennik, G.C. 1969. White clover/grass relationships: competition effects in laboratory and field. In *Proceedings White Clover Research Symposium.* Belfast, Sept. 1969. pp. 165–74.

Erbach, D.C. & W.G. Lovely. 1976. Machinery adaptations for multiple cropping. In Pappendick, P.A. Sanchez and G.B. Triplett (eds.) *Multiple Cropping.* ASA Sp. Pub. 27. pp. 337–46.

Faris, M.A., M.R.A. de Araujo, M. de A. Lira, & A.S.S. Arcovere. 1983. Yield stability in intercropping studies of sorghum or maize with cowpea or common beans under different fertility levels in northeastern Brazil. *Can. J. Plant Sci.* **63**:789–99.

Fawusi, M.O.A. 1985. Influence of spatial arrangements on the growth, fruit and grain yields and yield components of intercropped maize and okra (*Abelmoscus esculentus.*) *Field Crops Res.* **11**:345–52.

Federer. 1987. The statistical design and analysis of intercropping experiments. MS.

Feeny, P.P. 1976. Plant apparency and chemical defense. In J. Wallace and R. Mansel (eds.) *Biochemical Interaction Between Plants and Insects. Rec. Adv. Phytochem.* **10**:1–40.

Fisher, N.M. 1975. *Investigations into the Competitive Relations of Maize and Beans in Mixed Crops.* Technical Communication 14, Dept. of Crop Sci., University of Nairobi, Kenya.

Fisher, N.M. 1977*a*. Studies in mixed cropping. I. Seasonal differences in relative productivity of crop mixtures and pure stands in the Kenya Highlands. *Exp. Agrc.* **13**:177–84.

Fisher, N.M. 1977*b*. Studies in mixed cropping. II. Population pressures in maize-bean mixtures. *Exp. Agrc.* **13**:185–91.

Fleck, N.G., C.M.N. Machado & R.S. DeSouza. 1984. Effiéncia da consorciacao de culturas no controle de plantas daninhas. *Pesquisa Agropecuária Brasileira* **19**:591–8.

Francis, C.A. 1986*a*. Variety development for multiple cropping systems. *CRC Critical Rev. in Plant Sci.* **3**:133–68.

Francis, C.A. 1986*b*. *Multiple Cropping: Practices and Potentials.* Macmillan, New York.

Francis, C.A., C.A. Flor & M. Pragner. 1978. Effects of bean association on yields and yield components of maize. *Crop Sci.* **18**:760–4.

Francis, C.A., M. Prager, D.R. Laing & C.A. Flor. 1978. Genotype X Environment interactions in bush bean cultivars in monoculture and associated with maize. *Crop Sci.* **18**:237–46.

Francis, C.A. & J.H. Sanders. 1978. Economic analysis of bean and maize systems: monoculture versus associated cropping. *Field Crops Res.* **1**:319–35.

Friedland, W. & A. Barton. 1975. *Destalking the Wily Tomato: A Case Study in Social Consequences in California Agricultural Research.* Univer. of Calif. Davis: Dept. of Appl. Behav. Sciences Res. Mono. No 15.

Gamboa, W. & J.H. Vandermeer. 1988. Root and shoot interactions between *Phaseolus vulgaris* L. and *Cyperus rotundus.* In press, *Trop. Ecol.*

Garot, A. & Subadi. 1958. Coconut interplanted with cacao at Balong Estate, Java. *Warta Pusat pekebunan Negra* **8**:3–5.

Gause, G.F. 1934. *The Struggle for Existence.* Waverley Press, Baltimore, USA.

Gavarra, M.R. & R.S. Raros. 1975. Studies on the biology of the predatory wolf spider, *Lycosa pseudoannulata* Boes et Str (Aran: Lycosidae). *Phillippines Entomology* **2**:277–44.

Gerdeman, J.W. 1968. Vesicular-arbuscular mycorrhiza and plant growth. *Ann. Rev. Plant. Path.* **6**:397.

Giri, G. & R. De. 1978. Intercropping of pigeonpea with other grain legumes under semi-arid rainfed conditions. *Indian J. Ag. Sci.* **48**:659–65.

Gliemeroth, G. 1950. Untersuchungen über dis Einspritzung von Speiserbsen. *Seitschrift für Acker-und-Pflanzenbau* **91**:519–44.

Goldberg, D.E. 1982. The distribution of evergreen and deciduous trees relative to soil type: an example from the Sierra Madre, Mexico and a general model. *Ecology* **63**:942–51.

Goldberg, D.E. & P.A. Werner. 1983. Equivalence of competitors in plant communities: A null hypothesis and a field experimental approach. *Am J. Botany* **70**:1098–104.

Gompertz, B. 1825. On the nature of the function expressive of the law of human morality. *Phil. Trans. Roy. Soc. Lond.* **36**:513–85.

Goodall, D.W. 1960. Quantitative effects of intraspecific competition: an experiment with marigolds. *Bull. Res. Council of Israel* **8**:181–94.

Goodman, D. 1975. The theory of diversity-stability relationships in ecology. *Q. Rev. Biol.* **0**:237–66.

Govinden, N., J.T. Arnason, B.J.R. Philogene & J.D.H. Lambert. 1984. Intercropping in the tropics: advantages and relevance to the small farmer. *Canadian J. Dev. Studies* **5**:213–32.

Grant, P.R. 1972. convergent and divergent character displacement. *Biol. J. Linn. Soc.* **4**:39-68.

Greenwood, D.J. 1969. Effect of oxygen distribution in the soil on plant growth. In Whittingham, W.J. (ed.), *Root Growth. Proc. 15th Easter School Ag. Sci.* Nottingham. pp. 202–23.

Gregory, F.G. 1928. The analysis of growth curves – a reply to criticism. *Ann. Bot.* **42**:531–9.

Grime, J.P. 1979. *Plant Strategies and Vegetation Processes.* Wiley, New York.

Grimes, R.C. 1948. Intercropping and alternate row cropping of cotton and maize. *E. African Agricultural and Forestry J.* **28**:161–3.

Gunary, D. 1968. Discussion on mineral nutrient supply from soils. In H. Rorison (ed.), *Ecological Aspects of the Mineral Nutrition of Plants, Br. Ecol. Soc. Symp.* Blackwell, Oxford. pp. 149–52.

Gupta, S.L. 1953. The effect of mixed cropping of arhar (*Cajanus cajan* Srent) with jowar (*Sorghum vulgar* Pers) on the incidence of arhar cultivation. *Kanpur Agricultural College J.* **13**:18–25.

Gutierrez, V., M. Infante & A. Pinchinot. 1975. Situación del cultivo de frijol en America Latina. *Centro Internacional de Agricultura Tropical*, Cali, Colombia.

Haizel, K.A. & J.L. Harper. 1973. The effects of density and the timing of removal on interference between barley, white mustard and wild oats. *J. Appl. Ecol.* **10**:23–31.

Halbert, S.E. & M.E. Irwin. 1981. Effect of soybean canopy closure on landing rates of aphids with implications for restricting spread of soybean mosaic virus. *Ann. Appl. Biol.* **98**:15–19.

Hall, R.L. 1974. Analysis of the nature of interference between plants or different species. II. Nutrient relations in a Nandi Satoria and Greenleaf Desmodium association with particular reference to potassium. *Aust. J. Agric. Res.* **25**:749–56.

Hansen, M.K. 1983. Interactions among natural enemies, herbivores, and yield in monocultures and polycultures of corn, bean, and squash. Ph.D. Dissertation, University of Michigan, Ann Arbor.

Hardin, G. 1960. The competitive exclusion principle. *Science* **131**:1292–7.

Harper, J.L. 1977. *Population Biology of Plants.* Academic Press, London, p. 892.

Hawkins, R. 1984. Intercropping maize with sorghum in Central America: a cropping system case study. *Agric. Syst.* **15**:79–99.

Haynes, R.J. 1980. Competitive aspects of the grass-legume association. *Adv. in Agron.* **33**:227–61.

Haystead, A. 1983. The efficiency of utilization of biologically fixed nitrogen in crop production systems. In Jones, D.G. and D.R. Davies (eds.), *Temperate Legumes: Physiology, Genetics, and Modulation.* Pitman Adv. Publ. Program, Boston. pp. 395–415.

Hazlett, B.A. 1981. The behavioral ecology of hermit crabs. *Ann. Rev. Ecol. and Syst.* **76**:507–10.

Hazlett, B.A. 1983. Interspecific negotiations: Mutual gain in exchanges of a limiting resource, *Animal Behavior* **31**:160–3.

Herbert, S.J., S.H. Putnam, N.M. Poos-Floyd, A. Vargas & J.F. Creighton. 1984. Forage yield of intercropped corn and soybean in various planting patterns. *Agron. J.* **76**:507–10.

Holliday, R. 1960. Plant population and crop yield. *Field Crop Abstracts* **13**:159–67; 247–54.

Horn, H.S. 1971. *The Adaptive Geometry of Trees.* Monographs in Pop. Biol., Princeton University Press, Princeton, New Jersey.

Hsu, S.B., S.P. Hubbell & P. Waltman. A mathematical theory for single-nutrient competition in continuous cultures of microorganisms. *SIAM J. of Appl. Math.* **32**:366–83.

Hutchinson, G.E. 1957. Concluding remarks. *Cold Spring Harbor Symp. Quant. Biol.* **22**:415–27.

Huxley, P.A. (ed.). 1983. *Plant Research and Agroforestry*. Int. Council for Res. in Agroforestry (ICRAF), Nairobi, Kenya.

Huxley, P.A. & Z. Maingu. 1978. Use of a systematic spacing design as an aid to the study of intercropping: some general considerations. *Experimental Agric.* **14**:49–56.

Ibgozurike, M.V. 1971. Ecological balance in tropical agriculture. *Geographical Rev.* **61**:519–27.

Imle, E.P., A.L. Erickson & L.P. Oechsli. 1954. Performance of clonal cuttings and clonal seedlings of cacao interplanted with rubber. In *Reunión del comité Técnico Interamericano de Cacao*, Turrialba, Costa Rica. Trabajos presentados. Turrialba. Costa Rica. IICA. 1954. V.1.Sel.1.Dec.25.11 pp.

International Rice Research Institute. 1974. *IRRI Annual Report for 1973.* Los Banos, Philippines.

International Rice Research Institute. 1975. *IRRI Annual Report for 1974.* Los Banos, Philippines.

Jackson, J.E. 1983. Light climate and crop-tree mixtures. In Huxley, P.A. (ed.), *Plant Research and Agroforestry*. Int. Council for Res. in Agroforestry (ICRAF), Nairobi, Kenya.

Jahnke, L.S. & D.B. Lawrence. 1965. Influence of photosynthetic crown structure on potential productivity of vegetation, based primarily on mathematical models. *Ecology* **46**:319–26.

Janos, D.P. 1983. Tropical mycorrhizae, nutrient cycles and plant growth. In S.L. Sutton, T.C. Whitmore, and A.C. Chadwick (eds.), *Tropical Rain Forest: Ecology and Management.* Blackwell Sci. Publ., Oxford. pp. 327–45.

Jose, B.M. 1968. Intercropping cacao with coconut. *Coffee and Cacao J.* **11**:128–30.

Kareiva, P. 1982. Experimental and mathematical analyses of herbivore movement: quantifying the influence of plant spacing and quality on foraging discrimination. *Ecol. Monogr.* **52**:261–82.

Kareiva, P.M. 1983. Influence of vegetation texture on herbivore populations:

resource concentration and herbivore movement. In Denno, R.F. and M.S. McClure (eds.), *Variable Plants and Herbivores in Natural and Managed Systems.* Academic Press, New York, pp. 259–89.

Kareiva, P. 1986. Trivial movement and foraging by crop colonizers. In Kogan, M. (ed.), *Ecological Theory and IPM Practice.* Wiley, New York.

Kass, D.C. 1978. Polyculture cropping systems: review and analysis. *Cornell Int. Agr. Bull.* No. 32, 69 pp.

Kaurov, I.A. & T.A. Budkevitch. 1973. Kinetics of mineral nutrients in oats and peas in pure and mixed stands during growth and development. Vestsi Akademii Navak BSSR Biyagichnykh Navak 5. In *Field Crop Abstracts* 27; Abst. 1662. 1974.

Kawano, K. & M.D. Thung. 1982. Intergenotypic competition and competition with associated crops in cassava. *Crop Science* **22**:59–63.

Kellerman, K. & R.C. Wright. 1914. Mutual influence of certain crops in relation to nitrogen. *Agron. J.* **6**:204–10.

Khan, A.M., S.K. Saxena & Z.A. Siddiqui. 1971. Efficacy of *Tagetes erecta* in reducing root infesting nematodes of tomato and okra. *Indian Phytopathology* **24**:166–9.

Kowal, J. & D.F. Andrews. 1973. Pattern of water availability and water requirement for grain sorghum production at Samaru, Nigeria. *Trop. Agr. (Trinidad)*, **50**:89–100.

Kurtz, T., S.W. Melsted & R.H. Bray. 1952. The importance of nitrogen and water in reducing competition between intercrops and corn. *Agron. J.* **44**:13–17.

Lakhani, D.A. 1976. A crop physiological study of mixtures of sunflower and fodder radish. Ph.D. Thesis, Reading Univ. England 171 pp.

Lamberts, M.L. 1980. Intercropping with potatoes. MS Thesis, Cornell University.

Lawas, C.M. 1947. A study on the intercropping of corn with sweet potato. *University of Philippines, College of Agric. Bi-weekly Bull.* **12**:1–2.

Lawrence, W.S. 1987. Emigration patterns of the Red Milkweed Beetle from host plant patches: effects of conspecific abundance and sex ratio. *Ecology* (in press).

Leach, J.R. 1971. Underplanting coconuts with cacao in West Malaysia. *Cacao Grower's Bull.* **16**:21–6.

Lenga, F.K. & I. Stewart. 1982. Water-yield response of a maize and bean intercrop. *E. Afr. Agric. For. J.* **45**:103–16.

Leonard, K.J. 1969. Factors affecting rates of stem rust increase in mixed plantings of susceptible and resistant oat varieties. *Phytopathology* **59**:1845–50.

Levin, S.A. 1970. Community equilibria and stability, and an extension of the competitive exclusion principle. *Am. Nat.* **104**:413–23.

Levine, S.H. 1976. Competitive interactions in ecosystems. *Am. Nat.* **110**:903–10.

Levins, R. 1962. Theory of fitness in a heterogeneous environment. I. The fitness set and adaptive function. *Am. Nat.* **96**:361–73.

Levins, R. 1966. The strategy of model building in population biology. *Am. Scientist* **54**:421–31.

Levins, R. 1968. *Evolution in Changing Environments.* Princeton Monographs in Population Biology, Princeton Univ. Press. 120 pp.

Levins, R. 1975. Evolution in communities near equilibrium. In Cody, M.L. and J.M. Diamond (eds.) *Ecology and Evolution of Communities*, The Belknap Press of Harvard University Press, Cambridge, Mass., USA.

Levins, R. & R. Lewontin. 1980. Dialectics and reductionism in ecology. In *The Dialectical Biologist.* pp. 132–60. Harvard University Press, Cambridge, Mass., USA.

Lewontin, R.C. 1966. On the measurement of relative variability. *Sys. Zool.* **1**:141–2.

Leyton, L. 1983. Crop water use: principles and some considerations for agroforestry. In Huxley, P.A. (ed.), *Plant Research and Agroforestry.* Int. Council for Res. in Agroforestry, ICRAF. pp. 379–400.

Liboon, S.P. & R.R. Harwood. 1975. Nitrogen response in corn soybean intercropping. In *6th Annual Science Meeting (Proceedings), Philippines Crop Sci. Soc.*, Bacolod City, Philippines.

Liebman, M.Z. 1986. Ecological Suppression of Weeds in Intercropping Systems: Experiments with Barley, Pea, and Mustard. Ph.D. Dissertation, Berkeley, California.

Lincoln, C. & D. Isley. 1947. Corn as a trap crop for the cotton bollworm. *J. Econ. Ent.* **40**:437–8.

Lipman, J.G. 1912. The associative growth of legumes and non-legumes. *New Jersey Agr. Exp. Sta. Bul.* **253**. 48 pp.

Litsinger, J.A. & K. Moody. 1976. Integrated pest management in multiple cropping systems. In Papendick, R.I., P.A. Sanchez & G.B. Triplett (eds.) *Multiple Cropping.* ASA Special Publication No. 27. Am. Soc. Agron., Madison WI, USA, pp. 293–316.

Lolas, P.C. 1986. Weed community interference in burley and oriental tobacco (*Nicotiana tabacum*). *Weed Res.* **26**:1–7.

Loomis, R.S. 1984. Traditional agriculture in America. *Ann. Rev. Ecol. Syst.* **15**:449–78.

Lopez de la Fuente, J. 1986. *Red solar y la estación Vadstena Nicaragua* 1985. Reporte Témocp-Investigativo No. 04/86. UCA, Managua.

Lotka, A.J. 1925. *Principles of Physical Biology.* Waverly Press, Baltimore, USA.

Ludwig, W. 1950. Zur Theorie der Konkurrenz: die Annidation (Einnischung) als fünfter Evolutionsfaktor. *Zool. Anz. Ergänzungsband zu Band* **145**:516–37.

Lugo-Lopez, B.G. 1953. Intercropping sugarcane with food crops. *J. Agric.* Univ. Puerto Rico **37**:171–82.

MacArthur, R. 1972. *Geographical Ecology.* Harper and Row, New York.

MacArthur, R.H. & R. Levins. 1967. The limiting similarity, convergence and divergence of coexisting species. *Am. Nat.* **101**:377–8.

Maken, M.M. & Doto, A.L. 1982. Soybean-cereal intercropping and its implications in soybean breeding. In Keswani, C.L. and B.J. Ndunguru (eds.). *Intercropping Proc. of the Second Symp. on Intercropping in Semi-Arid Areas.* Morogoro, Tanzania. IDRC. Ottawa. pp. 84–90.

Marshall, D.R. & A.H.D. Brown. 1973. Stability of performance of mixtures and multilines. *Euphytica* **22**:405–12.

Martin, F.B. 1973. Beehive designs for observing variety competition. *Biometrics* **29**:397–402.

Martin, M.P.L.D., & R.W. Snaydon. 1982. Root and shoot interactions between barley and field beans when intercropped. *J. Appl. Biol.* **19**:263–72.
Martinez, J.T. 1947. A study of intercropping corn with cassava. *Coll. of Agric. Bull. (Philippines)* **12**:1–2.
May, R. 1974. *Stability and Complexity in Model Ecosystems.* Princeton Monog. in Pop. Biol., Princeton University Press, Princeton, New Jersey.
May, R.M. 1981. The role of theory in ecology. *Am Zool.* **21**:903–10.
May, R.M. & R.H. MacArthur. 1972. Niche overlap as a function of environmental variability. *Proc. Nat. Acad. Sci.*, U.S. **69**:1109–13.
McBeth, C.W. & A.L. Taylor. 1944. Immune and resistant cover crops valuable in rootknot infested peach orchards. *Proc. Am. Soc. Hort. Sci.* **45**:158–66.
McCowan, R.L. 1973. An evaluation of the influence of available water stoarage capacity on growth season length and yield of tropical pastures using simple water balance models. *Agr. Meteorol.* **11**:53–64.
McGuinness, H. 1987. The importance of plant diversity and the nutritional content of the diet on the population dynamics of herbivorous pest insects. Ph.D. Thesis, University of Michigan, Ann Arbor, MI.
McNaughton, S.J. 1977. Diversity and stability of ecological communities: a comment on the role of empiricism in ecology. *Am. Nat.* **111**:515–25.
Mead, R. 1979. Competition Experiments. *Biometrics* **35**:41–54.
Mead, R. & J. Riley. 1981. A review of statistical ideas relevant to intercropping research. *J. Royal Stat. Soc.* ser. A. **144**:462–509.
Mead, R. & R.D. Stern. 1981. Statistical considerations in experiments to investigate intercropping. In Willey, R. 1979. *Proceedings of the international workshop on intercropping.* ICRISAT (International Crops Research Institute for the Semi-Arid Tropics.)
Mead, R. & R.W. Willey. 1980. The concept of a 'Land Equivalent Ratio' and advantages in yields from intercropping. *Exp. Agr.* **16**:217–28.
Michon, G., J. Bompard, P. Hecketsweiler, & C. Ducatillion. 1983. Tropical forest architectural analysis as applied to agroforests in the humid tropics: the example of traditional village-agroforests in West Java. *Agroforestry Systems* **1**:117–29.
Monsi, M., & T. Saeki. 1953. Uber der Lichfaktor en den Pflanzengesellschaften und seine Bedeutung für die Stooffproduktion. *Jap., J. of Bot.* **14**:22–52.
Monteith, J.L. 1965. Light distribution and photosynthesis in field crops. *Ann. Bot.* **29**:17–37.
Moody, K. 1980. Weed control in intercropping in tropical Asia. In Akobundu, I.O. (ed.). *Weeds and their Control in the Humid and Subhumid Tropics.* Int. Inst. Trop. Agric., Ibadan, Nigeria. pp. 101–108.
Mooney, H.A. & S.L. Gulmon. 1979. Environmental and evolutionary constraints on the photosynthetic characteristics of higher plants. In O.T. Solbrig, S. Jain, G.B. Johnson and P.H. Raven (eds.). *Topics in plant population biology.* Columbia University Press. New York.
Moreno, R.A. & L.E. Mora. 1984. Cropping pattern and soil management influence on plant diseases: II. Bean rust epidemiology. *Turrialba* **34**:41–5.
Moursi, M.A. 1966. The interrelation between cotton and other crops grown

together. III. Effect of interplanting garlic with cotton plants on the chemical content and yield of cotton and garlic crops. *Ann. of Agric. Sci.* **11**:229–41.

Murashige, T. 1962. Papaya retards macademia growth. *Hawaii Farm Science* **11**:1–2.

Murdoch, W.W. 1975. Diversity, complexity, stability, and pest control. *J. Appl. Ecol.* **12**:795–807.

Nair, P.K.R. 1983. Agroforestry with coconuts and other tropical plantation crops. In Huxley, P.A. (ed.). *Plant Research and Agroforestry*. Int. Council for Res. in Agroforestry (ICRAF). Nairobi, Kenya. pp. 79–102.

Nair, P.K.R., U.K. Patel, R.P. Singh & M.K. Kaushik. 1979. Evaluation of legume intercropping in conservation of fertilizer nitrogen in maize culture. *J. Agric. Sci. Cambridge* **93**:189–94.

Newman, E.I. 1974. Root and soil water relations. In Carson, E.W. (ed.), *The Plant Root and its Environment*. University Press of Virginia, Charlottesville, VA. pp. 363–440.

Newman, S.M. 1986. A pear and vegetable interculture system: Land equivalent ratio, light use efficiency and productivity. *Exp. Agr.* **22**:383–92.

Nickel, J.L. 1973. Pest situation in changing agricultural systems – a review. *Bull. Entomol. Soc. Am.* **19**:136–42.

Nicol, H. 1934. The derivation of the nitrogen of crop plants with special reference to associated growth. *Biol. Rev.* **9**:383–410.

Nobbe, F. & L. Richter. 1902. Miteilungen aus der Königl. pflanzanphysiologischen Versuchs-Station. Therand. 58. Uber den Einfluss des Nitratstickstoffs und der Humussubstanzen auf den Impfungserfolg bei Leguminosen. *Landwirtschaftlichen Versuchs-Station* **56**:441–8.

Norman, D.W. 1974. Rationalizing mixed cropping under indigenous conditions: the example of northern Nigeria. *J. Devel. Stud.* **1**:3–21.

Norman, D.W. 1977. The rationalization of intercropping. *African Env.* **2/3**:97–109.

Norman, D.W., B.J. Buntjer, & A.D. Goddard. 1970. Intercropping observation plots at the farmers' level. *Samaru Agric. Newsletter* **12**:97–101.

Norman, M.J.T., C.J. Pearson & P.G.E. Searle. 1984. *The Ecology of Tropical Food Crops*. Cambridge University Press, Cambridge, England.

Nowotonowna, A. 1937. An investigation of nitrogen uptake in mixed crops not receiving nitrogenous manures. *J. Agr. Sci.* **27**:504–10.

Nye, P.H. 1968. The soil model and its application to plant nutrition. In Rorison, I.H. (ed.) *Ecological Aspects of the Mineral Nutrition of Plants*. 9th Symp. Br. Ecol. Soc., Blackwell, Oxford. pp. 105–14.

Oelsligle, D.D. & A.M. Pinchinat. 1975. Effect of varying nitrogen levels on grain yields, energy and protein production and economic returns of corn and beans when grown alone and in different combinations. Unpublished paper.

Okigbo, B.N. & R. Lal. 1978. Residue mulches and agrisilviculture in tropical African agriculture. Paper presented at International Conference on 'Basic Techniques in Ecological Agriculture' (2–5 Oct. Montreal, Canada. Cited in Steiner, K.G. 1984. Intercropping in Tropical Smallholder Agriculture with Special Reference to West Africa. Deutsche Gesellschaft für Technische Zusammenarbeit (GTZ).

Olosantan, F. 1985. Effects of intercropping, mulching and staking on growth and yield of tomatoes. *Exp. Agr.* **21**:135–44.

Olukosi, J.O. 1976. Decisions of farmers under risk and uncertainty. The case of Ipetu and Odo-Ore farmers in Kwara State. *Samaru Agric. Newsletter* **10**:108–22.

Osiru, D.S.O. & G.R. Kibira. 1979. Sorghum/pigeonpea and finger millet/groundnut mixtures with special reference to plant population and crop arrangement. In Willey, R.W. (ed.), *Proceedings of the International Workshop on Intercropping.* ICRISAT, pp. 78–85.

Osman, A.E. & A.M. Osman. 1982. Performance of mixtures of cereal and legume forages under irrigation in the Sudan. *J. Agric. Sci. Camb.* **98**:17–21.

Oyejola, B.A. & R. Mead. 1982. Statistical assessment of different ways of calculating land equivalent ratios (LER). *Exp. Agr.* **18**:125–38.

Paner, V.E. 1975. Multiple cropping research in the Philippines. In *Proc. of the cropping systems workshop,* IRRI, Los Banos, Philippines. pp. 188–202.

Papendick, R.I., P.A. Sanchez & G.B. Triplett. 1976. *Multiple Cropping.* ASA (American Society of Agronomy) Special publication no. 27.

Pearce, S.G. & R.N. Edmondson. 1982. Historical data as a guide to selecting systems for intercropping two species. *Exp. Agric.* **18**:353–62.

Pearce, S.G. & B. Gilliver. 1978. The statistical analysis of data from intercropping experiments. *J. Agr. Sci. Camb.* **91**:625–32.

Pearson, E.O. 1958. *The Insect Pests of Cotton in Tropical Africa.* Commonwealth Institute of Entomology, London. 355 pp.

Perfecto, I., B. Horwith, J. Vandermeer, B. Schultz, H. McGuinness & A. Dos Santos. 1986. Effects of plant diversity and density on the emigration rate of two ground beetles, *Harpalus pennsylvanicus* and *Evarthrus sodalis* (Coleoptera:Carabidae), in a system of tomatoes and beans. *Environ. Entomol.* **15**:1028–31.

Perrin, R.M. 1977. Pest management in multiple cropping systems. *Agroecosystems.* **3**:93–118.

Pfahler, P.L. 1965. Environmental variability and genetic diversity within populations of oats (cultured species of *Avena*) and rye (*Secale cereale* L.). *Crop Science* **5**:271–5.

Pianka, E.R. 1972. R and K-selection or b and d selection? *Am. Nat.* **106**:581–8.

Pianka, E.R. 1976. Competition and Niche Theory. pp. 114–41. In May, R. (ed.), *Theoretical Ecology: Principles and Applications.* Saunders, Philadelphia, USA.

Pielou, E.C. 1981. The usefulness of ecological models: a stock-taking. *Q. Rev. Biol.* **56**:17–31.

Pillai, M.R., A. Shanmugwasundoram, M. Govindarajan & S. Lal. 1957. Mixed cropping trials with ragi, cotton and groundnut. *Madras Agricultural J.* **44**:131–9.

Pilz, F. 1911. Leguminosen und graminen in Rein-und in Mengsaaten mit besonderer Berücksichtunge der Stickstoffnutzung. *Z. Landw. Versuchs Osterr.* **14**:1140–210.

Pinchinat, A.M., J. Soria & R. Bazan. 1976. Multiple Cropping in Tropical

America. pp. 51–61 in Papendick, R.I., P.A. Sanchez and G.B. Triplett (eds.), *Multiple Cropping.* Am. Soc. Agron. 27. pp. 51–61.

Plucknett, D.L. 1979. *Managing Pasture and Cattle under Coconuts.* Westview Press, Boulder, CO, USA.

Power, A. 1987. Plant community diversity, herbivore movement, and an insect-transmitted disease of tropical maize. *Ecology* (in press).

Punzalan, F.L. 1972. Field screening of herbicides for weed control in corn intercropped with peanut. pp. 87–90 in *Weed Science Report* 1971–72. Laguna, Philippines: Dep. of Agricultural Botany, University of Philippines.

Pushparajah, E. & S.V. Tan. 1970. Tapioca as an intercrop in rubber. In E.K. Blencowe and J.W. Blencowe (eds.), *Crop Diversification in Malaysia,* Kuala Lumpur.

Putnam, D.H., S.J. Herbert & A. Vargas. 1985. Intercropped corn-soya bean density studies: Yield complementarity. *Exp. Agr.* **21**:41–51.

Quesada, F., E. Somarriba, & E. Vargas. 1987. Modelo para la simulación de patrones de sombra de árboles. *CATIE, Serie Técnica, Informe Tecnicoll* **8**:1–91.

Radke, J.K. & W.C. Burrows. 1970. Soybean plant response to temporary field Windbreaks. *Agron. J.* **62**:424–9.

Radke, J.K. & R.T. Hagstrom. 1974. Wind turbulence in a soybean field sheltered by four types of wind barriers. *Agron. J.* **66**:273–8.

Radke, J.K. & R.T. Hagstrom. 1976. Strip Intercropping for Wind Protection. In Papendick, R.I., P.A. Sanchez and G.B. Triplett (eds.), *Multiple Cropping.* Am. Soc. Agron., Spec. Pub. 27. pp. 201–22.

Raintree, J.B. 1983. Bioeconomic considerations in the design of agroforestry cropping systems. In Huxley, P.A. (ed.), *Plant Research and Agroforestry.* Int. Council Res. in Agroforestry, ICRAF. pp. 271–89.

Rao, M.R. & R.W. Willey. 1980*a*. Preliminary studies of intercropping combinations based on pigeonpea or sorghum. *Exp. Agr.* **16**:29–39.

Rao, M.R. & R.W. Willey. 1980*b*. Evaluation of yield stability in intercropping: studies on sorghum/pigeonpea. *Exp. Agr.* **16**:105–16.

Rao, M.R. & R.W. Willey. 1980*c*. Preliminary studies of intercropping combinations based on pigeonpea or sorghum. *Exp. Agr.* **16**:29–39.

Rao, M.R. & R.W. Willey. 1981. Stability performance of a pigeonpea/sorghum intercrop system. In R.W. Wiley (ed.), *Proceedings of the International Workshop on Intercropping.* ICRISAT, Hyderabad, India. pp. 306–17.

Rao, M.R. & R.W. Willey. 1983. Effects of genotype in cereal/pigeonpea intercropping on the alfisols of the semi-arid tropics of India. *Exp. Agr.* **19**:67–78.

Rao, N.G.P., B.S. Rana & P.P. Tarhalkar. 1981. Stability productivity, and profitability of some intercropping systems in dryland agriculture. In Willey, R.W. (ed.), *Proceedings of the International Workshop on Intercropping.* ICRISAT, Hyderabad, India. pp. 292–8.

Rappaport, R.A. 1967. *Pigs for the Ancestors.* Yale University Press, New Haven.

Rappaport, R.A. 1971. The flow of energy in an agricultural society. *Sci. Amer.* **225**:116–32.

Rasmussen, W.D. 1978. Advances in American Agriculture: The mechanical tomato harvester as a case study. *Technology and Culture* **9**:531–43.

Rathcke, B. 1984. Competition and facilitation among plants for pollination. In Real, L. (ed.), *Pollination Biology*. Academic Press, New York.

Readhead, J.F., J.A. Maghembe & B.J. Ndunguru. 1983. The intercropping of grain legumes in agroforestry systems. In Huxley, P.A. (ed.), *Plant Research and Agroforestry*. Int. Council for Res. in Agroforestry (ICRAF), Nairobi, Kenya.

Reddy, M.S. & R.W. Willey. 1979. Evaluation of alternate cropping systems for alfisols of the Indian semi-arid tropics. *Exp. Agr.* **21**:271–80.

Reddy, M.S. & R.W. Willey. 1981. A study of pearl millet/groundnut intercropping with particular emphasis on the efficiencies of leaf canopy and rooting pattern. In Willey (ed.), *Proceedings of the International Workshop on Intercropping*. ICRSAT, Hyderabad, India.

Reddy, N.N. & B.N. Chatterjee. 1973. Intercropping of soybean and rice. *Indian J. Agron.* **18**:464–72.

Reed, H.S. 1920. The nature of the growth rate. *J. Gen. Phys.* **2**:545–61.

Reed, H.S. & R.H. Holland. 1919. The growth rate of an annual plant *Helianthus. Proc. Nat. Acad. Sci. Wash.* **5**:135–44.

Reich, V.R. & R.E. Atkins. 1970. Yield stability of four population types of grain sorghum, *Sorghum bicolor* (L.) Moench, in different environments. *Crop Science* **10**:11–17.

Rerkasem, K., W.R. Stern & N.A. Goodchild. 1980. Associated growth of wheat and ryegrass. I. Effect of varying total density and proportion in mixtures of wheat and annual ryegrass. *Aust. J. Ag. Res.* **31**:649–58.

Rheenen, Van H.A., O.E. Hasselbach & S.G.S. Muigai. 1981. The effect of growing beans together with maize on the incidence of bean diseases and pests. *Neth. J. Pl. Path.* **7**:193–9.

Richards, F.J. 1959. A flexible growth function for empirical use. *J. Exp. Bot.* **10**:290–300.

Riley, J. 1985. Examination of the staple and effective land equivalent ratios. *Exp. Agr.* **21**:369–76.

Risch, S.J. 1980. The population dynamics of several herbivorous beetles in a tropical agroecosystem: the effect of intercropping corn, beans and squash in Costa Rica. *J. Appl. Ecol.* **17**:593–612.

Risch, S.J. 1981. Insect herbivore abundance in tropical monocultures and polycultures: an experimental test of two hypotheses. *Ecology* **62**:1325–40.

Risch, S.J., D. Andow, & M.A. Altieri. 1983. Agroecosystem diversity and pest control: Data, tentative conclusions, and new research directions. *Environ. Entomol.* **12**:625–9.

Robertson, T.B. 1923. *The Chemical Basis of Growth and Senescence.* Loppincott, Philadelphia and London.

Robinson, J.B.D. 1962. The influence of interplanted bananas on Arabica coffee yields. In *Tanganyika, Coffee Res. Stat., Lyamungu. Res. Rept. for* 1961. Dar-es-Salaam pp. 31–8.

Robinson, R.C. 1960. Oat-pea and oat-vetch mixtures for forage and seed. *Agron. J.* **52**:546–9.

Robinson, R.C. & R.J. Dunham. 1954. Companion crops for weed control in soybeans. *Agron. J.* **54**:278–81.

Room, P.M. 1972. The fauna of the mistletoe *Tapinanthus banquensis* growing on cocoa in Ghana relationships between fauna and mistletoe. *J. Animal Ecol.* **43**:95–124.

Root, R. 1973. Organization of a plant-arthropod association in simple and diverse habitats The fauna of collards (*Brassica oleracea*). *Ecol. Monogr.* **43**:95–124.

Rosenberg, N.J. 1966. Influence of snow fence and corn windbreaks on microclimate and growth of irrigated sugar beets. *Agron. J.* **58**:469–75.

Rosset, P.M. 1986. Ecological and economic aspects of pest management and polycultures of tomatoes in Central America. Ph.D. Dissertation, University of Michigan, Ann Arbor, Michigan, USA.

Rosset, P.M., R.J. Ambrose, A.G. Power & A. Hruska. 1984. Overyielding in a polyculture of tomatoes and beans in Costa Rica. *Tropical Agric.* (*Trinidad*) **61**:208–12.

Rosset, P.M. & J.H. Vandermeer. 1986. The confrontation between processors and farmworkers in the midwest tomato industry and the role of the agricultural research and extension establishment. *Agric. and Human Values*, **3**:26–32.

Rosset, P.M., J. Vandermeer, M. Cano, G. Varrela, A. Snook & C. Hellpap. 1986. El frijol como cultivo trampa para el combate de *Spodoptera sunia* Guenèe (Lepidopteraa:Noctuidae) en plantulas de tomate. *Agronomia Costaricense*, **9**:99–102.

Roughgarden, J. 1972. Evolution of niche width. *Am. Nat.* **106**:683–718.

Roughgarden, J. 1974. Species packing and the competition function with illustrations from coral reef fish. *Theoret. Pop. Biol.* **5**:163–86.

Roughgarden, J. 1983. Competition and theory in community ecology. *Am. Nat.* **122**:583–601.

Rouillard, G. & G. Mazery. 1969. Notes on sunflower cultivation in ratoon cane interlines. In *Mauritius Sugar Industry Res. Inst. Ann. Rept.*, 1968.

Russell, W.M.S. 1968. The slash-and-burn Technique. *Nat. Hist.* **78**:58–65.

Saeki, T. 1960. Interrelationships between leaf amount, light distribution, and total photosynthesis in a plant community. *Botanical Magazine, Tokyo*, **73**:404–8.

Sanchez, P.A. 1976. Properties and Management of Soils in the tropics. Wiley, New York, pp. 478–532.

Santhirasegaram, K. 1967. Intercropping of coconuts with special reference to food production. *Ceylon coconut Planters' Rev.* **5**:12–24.

Sarup, P., V.K. Sharma, V.P.S. Panwar, K.H. Siddiqui, K.H. Marwaha & K.N. Agarwal. 1977. Economic threshold of *Chilo partellus* infesting maize crop. *J. Ent. Res.* **1**:92–9.

Schilling, R. 1965. L'Arachide en cultures associées avec les cereales. *Oleagineux* **20**:673–6.

Schoener, T.W. 1965. The evolution of bill size differences among sympatric cogeneric species of birds. *Evolution* **19**:189–213.

Schoener, T.W. 1968. The Anolis lizards of Bimini resource partitioning in a complex fauna. *Ecology* **49**:704–26.

Schoener, T.W. 1974. Resource partitioning in ecological communities. *Science* **185**:27–39.

Schultz, B.B. 1984. Ecological aspects of stability in polycultures versus sets of monocultures of annual crops. Ph.D. Dissertation, University of Michigan, Ann Arbor, Michigan, USA.

Schultz, B.B., C. Phillips, P. Rosset, & J. Vandermeer. 1982. An experiment in intercropping tomatoes and cucumbers in southern Michigan, USA. *Scientia Horticulturae* **18**:1–8.

Scott, W.O. & F.L. Patterson. 1962. Grain sorghum as a companion crop for alfalfa. *Agron. J.* **54**:253–6.

Senenayake, Y.D.A. 1968. Intercropping, supplementary cropping and crop substitution on rubber land; a viewpoint. *Bull. Rubber Res. Inst.* **3**:99–113.

Shackel, K.A. & A.E. Hall. 1984. Effect of intercropping on the water relations of sorghum and cowpea. *Field Crops Res.* **8**:381–7.

Sharma, D., Lingh Laxman, & S.K. Maheswari. 1973. In M.P., soybean-arhar ensures more profit. *Indian Farming* **23**:33.

Shetty, S.V.R. & A.N. Rao. 1979. Weed-management studies in sorghum/pigeonpea and pearl millet/groundnut intercrop systems – some observations. In Willey, R.W. (ed.), *Proceedings of the International Workshop on Intercropping.* ICRISAT, Hyderabad, India. pp. 238–48.

Shia, F.Y. & T.P. Pao. 1964. On the yields of sugar-cane interplanted with different varieties of sweet potato (in Chinese). *Report of the Taiwan Sugar Expt. St.* **35**:55–63.

Shinozaki, K. & T. Kira. 1956. Intraspecific competition among higher plants. VII. Logistic theory of the C-D effect. *J. Inst. Polytech. Osaka Cy. Univ.* **7**:3–72.

Short, S.R.H. and D.W. Kretchman. 1974. Windbreaks for direct-seeded tomatoes. *Res. Summ. Ohio Agric. Res. Dev. Center* **72**:7–8.

Silvertown, J.W. 1982. *Introduction to Plant Population Ecology*. Longman, London, 209 pp.

Simberloff, D.S. 1983. Competition theory, hypothesis testing, and other community ecological buzzwords. *Am. Nat.* **122**:626–35.

Simpson, J.R. 1965. The transference of nitrogen from pasture legumes associated grass under several systems of management in pot culture. *Aust. J. Agric. Res.* **16**:915–26.

Singh, J.N., D.S. Negi & S.K. Tripathi. 1973. Study on intercropping of soybean with maize and jowar, *Indian J. Agron.* **18**:75–8.

Singh, N.B., P.P. Singh & K.P.P. Nair. 1986. Effect of legume intercropping on enrichment of soil nitrogen, bacterial activity and productivity of associated maize crops. *Exp. Agr.* **22**:339–44.

Singh, R.D. & P. Chand. 1969. Intercropping of maize with forage legumes. *Indian J. Agron.* **14**:67–70.

Slatkin, M. & D.J. Anderson. 1984. A model for competition for space. *Ecology* **65**:1840–5.

Smith, J.G. 1969. Some effects of crop background on populations of aphids and their natural enemies on Brussels sprouts. *Ann. Appl. Biol.* **63**:326–9.

Snaydon, R.W. 1971. An analysis of competition between plants of *Trifolium repens* L., populations collected from contrasting soils. *J. Appl. Ecol.* **8**:687–97.

Snaydon, R.W. 1979. A new technique for studying plant interactions. *J. Appl. Ecol.* **16**:281–6.

Snaydon, R.W. & P.M. Harris. 1979. Interactions belowground – the use of nutrients and water. In Willey, R.W. (ed.), *Proceedings of the International Workshop on Intercropping.* ICRISAT, Hyderabad, India. pp. 188–201.

Soekarno, T. 1961. A report on cacao in Indonesia. *Coffee and Cacao J.* **4**:64–5.

Soria, J.R. Bazan, A.M. Pinchinat, G. Paez, N. Mateo, R. Moreno, J. Fargas, & W. Forsythe. 1975. Investigación sobre sistemas de producción agricola para el pequeno agricultor del trópico. *Turrialba* **25**:283–93.

Southwood, T.R.E. 1978. *Ecological Methods.* Chapman and Hall, London, 524 pp.

Sparnaaij, L.D. 1957. Mixed cropping in oil palm cultivation. *J. West African Inst. of Oil Palm Res.* **2**:244–64.

Speight, M.R. 1983. The potential of ecosystem management for pest control. *Agriculture, Ecosystems and Environment* **10**:183–99.

Speight, M.R. & J.H. Lawton. 1976. The influence of weed-cover on the mortality imposed on artificial prey by predatory ground beetles in cereal fields. *Oecologia (Berl.)* **23**:211–23.

Spencer, J.E. 1966. *Shifting Cultivation in Southeastern Asia.* University of California Press, Berkeley, 247 pp.

Spurr, S.H. & B.V. Barnes. 1980. *Forest Ecology.* Wiley, New York.

Srivastava, H.P. 1972. Multiple cropping programme – progress and problems, pp. 335–42. In *Multiple Cropping Proc. of a Symp. Harayna Agric. Univ. Hissar.*

Stallings, J.H. 1926. The form of legume nitrogen assimilated by non-legumes when grown in association. *Soil Sci.* **21**:253–76.

Stanhill, G. 1962. The effect of environmental factors on the growth of alfalfa in the field. *Netherlands J. of Ag. Sci.* **10**:247–53.

Stanton, M.L. 1983. Spatial patterns in the plant community and their effects upon insect search. In Ahmad, S. (ed.), *Herbivorous Insects; Host-seeking Behavior and Mechanisms.* Academic Press, New York. pp. 125–56.

Steiner, K.G. 1984. *Intercropping in Tropical Smallholder Agriculture with Special Reference to West Africa.* Deutsche Gesellschaft für Technische Zusammenarbeit (GTZ), Eschborn, 304 pp.

Stern, W.R. & C.M. Donald. 1962. Light relationship in grass-clover sward. *Aust. J. Ag. Res.* **13**:599–614.

Stetner, R. 1976. The effects of mixed croppings in reduction of disease and insect damage in the tropics and subtropics. Term paper, Cornell University. 8 pp.

Sung, C.H. & I.K. Wu. 1966. Effects of interplanting tobacco in rice on characters of rice (In Japanese). In *Taiwan Tobacco and Wine Monopol Bur., Report for* 1966. pp. 45–50.

Tariah, N.M. & T.A.T. Wahua. 1985. Effects of component populations on yields and land equivalent ratios of intercropped maize and cowpea. *Field Crps. Res.* **12**:81–9.

Thatcher, L.E. 1925. The soybean in Ohio. *Ohio Agr. Exp. Sta. Bul.* 384.

Tilman, D. 1982. *Resource competition and community structure.* Monog. in Pop. Biol. Princeton University Press, Princeton, New Jersey.

Tilman, D. 1984. Plant dominance along an experimental nutrient gradient. *Ecology* **65**:1445–53.

Tilman, D. 1985. The resource-ratio hypothesis of plant succession. *Am. Nat.* **12**:827–52.

Tilman, D. 1986. Evolution and differentiation in terrestrial plant communities: the importance of the soil resource: light gradient. In Diamond J. and T.J. Case (eds.), *Community Ecology*. Harper and Row, New York. pp. 359–80.

Townsend, Jr., C.H., M.E. Espinosa & D. Fiester. 1964. Cultivo de hule (*Hevea brasiliensis*) intercalado en cafetales. *Guatemala Inst. Agropecuario Naccional. Boletin Tecnico* **6**:1–6.

Traeholt, P. 1962. The cacao industry in Malaya, ways of introducing it and its prospects. *Planter* **38**:248–51.

Trenbath, B.R. 1974. Biomass productivity of mixtures. *Adv. Agron.* **26**:177–210.

Trenbath, B.R. 1976. Plant interactions in mixed crop communities. In Papendick, R.I., P.A. Sanchez, and G.B. Triplett (eds.), *Multiple Cropping*. ASA special publication no. 27. Amer. Soc. Agron., Madison, WI, USA. pp. 129–70.

Trenbath, B.R. 1981. Light-use efficiency of crops and the potential for improvement through intercropping. In Willey, R.W. (ed.), *Proceedings of the International Workshop on Intercropping*. ICRISAT, Hyderabad, India. pp. 141–54.

Unamma, R.P.A. & L.S.O. Ene. 1983. Weed interference in cassava-maize intercrop in the rainforest of Nigeria. In *Trop. Root Crops. Prod. and uses in Africa. Proc. Second Triennial Symp. Int. Soc. Trop. Root Crops* – Africa Branch IDRC. pp. 59–62.

Unamma, R.P.A., L.S.O. Ene, S.O. Odurkwe & T. Enyinnia. 1986. Integrated weed management for cassava intercropped with maize. *Weed Res.* **26**:9–17.

Usher, M.B. 1970. An algorithm for estimating the length and direction of shadows with reference to the shadows of shelter belts. *J. Appl. Ecol.* **7**:141–5.

van den Berg, J.P. 1968. An analysis of yields of grasses in mixed and pure stands. *Versl. Landbouwkd Onderz* **714**:1–71.

van Huis, A. 1981. Integrated pest management in the small farmer's maize crop in Nicaragua. *Mededelingen Landbouwhogeschool, Wageningen*. pp. 1–221.

van Kesse, C., P.W. Singleton & H.J. Hoben. 1985. Enhanced N-transfer from soybean to maize by vesicular arbuscular mycorrhizal (VAM) fungi. *Plant Physiol.* **79**:562–3.

Vandermeer, J.H. 1970. The community matrix and the number of species in a community. *Am. Nat.* **104**:73–83.

Vandermeer, J.H. 1972. Niche theory. *Ann. Rev. Ecol. Syst.* **3**:107–32.

Vandermeer, J.H. 1980a. Saguaros and nurse trees: a new hypothesis to account for population fluctuations. *Southwestern Naturalist* **25**:357–60.

Vandermeer, J.H. 1980b. Indirect mutualism: variations on a theme by Stephen Levine. *Am. Nat.* **116**:441–8.

Vandermeer, J.H. 1981*a*. The interference production principle: an ecological theory for agriculture. *Bioscience* **31**:361–4.

Vandermeer, J.H. 1981*b*. Agricultural research and social conflict. *Science for the People*. **135**:5–8, 25–9.
Vandermeer, J.H. 1983. Una teoria de siembra intercalada en plantaciones jóvenes. *Ciencias de la Agricultura* (*Havana*) **15**:117–23.
Vandermeer, J.H. 1984*a*. Plant competition and the yield-density relation. *J. Theo. Biol*. **109**:393–9.
Vandermeer, J.H. 1984*b*. The interpretation and design of intercrop systems involving environmental modification by one of the components: A theoretical framework. *Bio. Agr. Hort*. **2**:135–56.
Vandermeer, J.H. 1986*a*. A computer-based technique for rapidly screening intercropping designs. *Exp. Agr*. **22**:215–32.
Vandermeer, J.H. 1986*b*. Mechanized agriculture and social welfare: the tomato harvester in Ohio. *Ag. and Human Values* **3**:21–5.
Vandermeer, J.H., R.J. Ambrose, M.K. Hansen, H. McGuinness, I. Perfecto, C. Phillips, P. Rosset & B. Schultz. 1984. An ecologically-based approach to the design of intercrop agroecosystems an intercropping system of soybeans and tomatoes in southern Michigan. *Ecological Modelling* **25**:121–50.
Vandermeer, J.H. & D.A. Andow. 1986. Prophylactic and responsive components of an integrated pest management program. *J. Econ. Entomol*. **79**:299–302.
Vandermeer, J.H., S. Gliessman, K. Yih & M. Amador. 1983. Overyielding in a corn-cowpea system in southern Mexico. *Biol. Agr. Hort*. **1**:83–96.
Vandermeer, J.H., B. Hazlett & B. Rathcke. 1985. Indirect facilitation and mutualism. In Boucher, D. (ed.), *The Biology of Mutualism*. Croom Helm, London. pp. 326–43.
Vandermeer, J.H. & A. Meyrat. 1988. La sombra y su calidad bajo una plantación de arboles: el potencial para sembrar un segundo cultivo y el diseno optimo para la plantación. MS.
Veevers, A. & T.B. Boffey. 1975. On the existence of leveled beehive designs. *Biometrics* **31**:963–7.
Vincent, J.M. 1974. Root-nodule symbioses with *Rhizobium*. In Quispe, A. (ed.), *The Biology of Nitrogen Fixation*. Amer. Elsevier, New York. pp. 265–341.
Virtanen, A.I., S. von Hausen, & T. Laine. 1937. Investigations on the root nodule bacteria of leguminous plants. Excretion of nitrogen in associated cultures of legumes and non-legumes. *J. Agric. Sci*. **27**:584–611.
Von Bertalanfy, L. 1957. Quantitative laws for metabolism and growth. *Quart. Rev. Biol*. **31**:217–31.
Wagmare, A.B. & S.P. Singh. 1984. Sorghum-legume intercropping and the effects of nitrogen fertilization. I. Yield and nitrogen uptake by crops. *Exp. Agr*. **20**:251–9.
Walker, T.W., H.D. Orchiston & A.R.F. Adams. 1954. The nitrogen economy of grass legume associations. *J. Brit. Grassl. Soc*. **9**:249.
Warren Wilson, J. 1961. Influence of spatial arrangement of foliage area on light interception and pasture growth. In *Proceedings*, *8th International grassland Congress*, 1960, Reading, England. pp. 275–9.

Wartiovaara, U. 1933. Uber den Stickstoffsaushalt des Hafers bei feld Mässigen Mischkulturen zusammen mit der Erbse. *Z. Pflanenernährung Düngung* **31**:353–9.

Watkinson, A.R. 1980. Density-dependence in a single species population of plants. *J. Theor. Biol.* **83**:345–57.

Watson, G.A. 1983. Development of mixed tree and food crop systems in the humid tropics: a response to population pressure and de-forestation. *Exp. Agr.* **19**:311–32.

Watt, A.S. & Fraser, G.K. 1933. Tree roots and the field layer. *J. Ecol.* **21**:404–14.

Webster, C.C. 1969. Notes on intercropping. In *Malaysian Oil Palm Conference, 2nd Kuala Lumpur. Progress in Oil Palm Research*, pp. 230–37.

Weiner, J. & S.C. Thomas. 1986. Size variability and competition in plant monocultures. *Oikos* **47**:211–22.

Werner, P.A. 1987. The effects of vegetation and edaphic gradients on goldenrods (*Solidago* spp.) in virgin prairie and old-field habitats: a field experiment using clonal reciprocal transplants. *Ecol. Monogr.* in press.

Westoby, M. 1984. The self-thinning rule. *Adv. in Ecol. Res.* **14**:167–225.

White, J. 1981. The allometric interpretation of the self-thinning rule. *J. Theor. Biol.* **89**:475–500.

White, J. & J. Harper. 1970. Correlated changes in plant size and number of plant populations. *J. Ecol.* **8**:467–48.

Whitehead, D.C. 1970. Uptake and assimilation of nitrogen by grass. In: *The Role of Nitrogen in Grassland Productivity. Commonw. Bur. Pastures and Field Crops. Bull.* No. 48:4–18. Hurle, Berkshire, England.

Wijesinha, A., A.T. Federer, J.R.P. Carvalho & T. de Aquino Portes. 1982. Some statistical analyses for a maize and beans intercropping experiment. *Crop Sci.* **22**:660–6.

Willey, R.W. 1979*a*. Intercropping – its importance and its research needs. Part I. Competition and yield advantages. *Field Crop Abstracts* **32**:1–10.

Willey, R.W. 1979*b*. Intercropping – its importance and its research needs. Part II. Agronomic relationships. *Field Crop Abstracts* **32**:73–85.

Willey, R.W. 1981. *Proceedings of the International Workshop on Intercropping*, ICRISAT, Hyderabad, India, 401 pp.

Willey, R. 1985. Evaluation and presentation of intercropping advantages. *Exp. Agr.* **21**:119–33.

Willey, R.W. & S.B. Heath. 1969. The quantitative relationships between plant population and crop yield. *Adv. Agron.* **21**:281–321.

Willey, R.W. & D.S.O. Osiru. 1972. Studies on mixtures of maize and beans (*Phaseolus vulgaris*) with particular reference to plant population. *J. Agr. Sci Cambridge* **79**:519–29.

Wit, C.T. de, P.G. Tow & G.C. Ennik. 1966. Competition between legumes and grasses. *Versl. Landbouwk. Onderz.* **687**:3–30.

Wood, G.A.R. 1966. A note on interplanting oil palms with cacao. *Planter* **42**:555.
Yoda, K., T. Kira, H. Ogawa & K. Hozumi. 1963. Self-thinning in over-crowded pure stands under cultivation and natural conditions. *J. Biol. Osaka City Univ.* **14**:107–29.

Author index

Subject index